Events and Festivals

Events and Festivals have an increasingly vital role in our leisure lifestyles. Indeed for many people events and festivals play a very significant part in their whole life.

The network of festivals and events that adorn the world most often serve to motivate visits as well as enhance the lives of the people who live in or near the host area. They are also dynamos of cultural development, knowledge and excellence. Whether in the area of sport, or culture and arts they can synthesise sophistication and enhance pleasure. Such dynamic outputs require dynamic inputs. This book looks at both forces through event and festival case examples – in a variety of international contexts, with each author or authors applying a range of current and new managerial and theoretical responses to their respective example.

Both up-to-date and forwarding looking, the managerial themes addressed are: Creative Management, Festival Audience Development, Culture and Community, Event and Festival Evaluation, Mentoring Volunteers, Time Management, and Festival Manager Perception. The festival and event types included are sport events, art festivals, community events, live music and culinary extravaganza.

This book was previously published as a special issue of *Managing Leisure: An International Journal*.

Martin Robertson is a lecturer in Tourism Management and a committee member of the Centre of Festival and Event Management in the School of Marketing, Tourism and Languages, Napier University Business School, Edinburgh. Martin's research is applied to sport events and festivals, with particular focus on the areas of brand narrative, and the analysis of economic and social impact.

Elspeth Frew is a senior lecturer in Tourism Management in the School of Sport, Tourism and Hospitality Management, LaTrobe University, Australia. Elspeth's research is in cultural tourism, with particular focus on festival and attraction management, industrial tourism and personality and tourism.

Events and Festivals

Current Trends and Issues

Edited by
Martin Robertson and Elspeth Frew

Routledge
Taylor & Francis Group

LONDON AND NEW YORK

First published 2008 by Routledge
2 Park Square, Milton Park, Abingdon, Oxon, OX14 4RN

Simultaneously published in the USA and Canada
by Routledge
270 Madison Avenue, New York, NY 10016

Routledge is an imprint of the Taylor & Francis Group, an informa business

2008 Taylor & Francis

Typeset in Cheltenham-Book by Techset Composition, Salisbury, UK
Printed and bound in Great Britain by MPG Books Ltd, Bodmin, Cornwall

British Library Cataloguing in Publication Data
A catalogue record for this book is available from the British Library

ISBN 10: 0-415-44918-9 (hbk)
ISBN 13: 978-0-415-44918-2 (hbk)

Contents

Events and festivals: Current trends and issues

Martin Robertson, Dr Donna Chambers and Dr Elspeth Frew

INTRODUCTION

Festival and events are prime manifestations of the experiencing economy (Pine and Gilmour, 1999), engaging memory, stimulating entertainment and acting as a dynamic for many other outcomes. From a myriad of societal requirements, *e.g.* economic, social, cultural, political, spiritual there may be a any number of outcomes – some anticipated, some not. Uniqueness, repetition, commemoration, image, risk, passion, knowledge and business are words that also appear in conjunction with festival and events documentation, both academic and in the wider social realm. Getz (2002) and Silvers, Bowdin, O'Toole and Nelson (2006) have both highlighted the complexity of the theoretical, methodological and ethical debate that interest in events and festivals have ignited. Both forward frameworks for the purpose of safeguarding the academic and professional legitimacy of the subject area.

The goal of this special issue, then, is to present some of the theoretical, applied and cross-disciplinary discourse and analysis, which has served to rouse increased interest in the festival and event subject area. Barlow and Shibli start with an evaluation of the role of *Music in the Rounds* touring classical music event, and a variety of marketing initiatives, in stimulating the development of both audience figures and the audience participation trends. Concentrating on the case of chamber music Barlow and Shibli give positive, if cautionary, conclusions about the role of the activities instigated by Arts Council, England.

Analysing a very different type of music event Beaven and Laws employ a multi-methodological model to evaluate the relationship between performances, internet ticket distribution channel platforms and the event audiences (fans). Concentrating on the tour by the UK popular music band Depeche Mode 2005 to 2006 tour in the United States, Beaven and Laws indicate some of the failings of the ticket distribution system pose as a predictive tool and as an appropriate media for engaging a fan base.

In comparing the policy basis of Glasgow's Winter Festival with the actual outcome – both reported and through application of a conceptual model engaging social, cultural and economic review – Foley and McPherson have produced a systematic and evaluative narrative. While focussed on the various geographical and cultural community considerations that Glasgow alone holds, and thus specific to this city, it does nonetheless draw findings and conclusions pertinent to the various other cities currently offering similar events with similarly stated objectives.

Hede and Rentschler have engaged in an evaluative study of a mentoring system employed for volunteer arts festival managers in the state of Victoria, Australia. Based on six case studies, the article addresses the need for mentoring systems in the arts (event) arena, which are goal orientated and not just based on knowledge and traditional competencies. The mentoring system *Directions* offers a number of innovative processes which the article investigates utilizing the Wolf (2004) five-stage approach for evaluation.

Definitive evaluation of visitor and economic impacts for events and festivals in the rural environment are far fewer than those undertaken in cities. Çela, Knowles-Lankford and Lankford, however, forward a comprehensive record and investigation of the impacts of a local food festival in north east Iowa, USA. The authors also suggest some significant future directions for research in progressing benefits of events to match the rural community composition of the area. They highlight the role of such events in creating opportunities for networks of information, partnerships and sophisticated cluster models to enable more strategic marketing, and the viability of more sustainable tourism development in which the needs of agriculture producers and visitor are complimentary.

There is increasing recognition by both theorists and art festival managers that the needs of stakeholders should form at least part of the measurement of success of these events. Recognizing that there are few records – and certainly not comparative ones – Williams and Bowdin offer a valuable compilation, from seven art festivals in the UK, of the evaluation methods employed, data types produced and the tools and techniques used in collecting and formatting results. The authors forward a level of evaluation rating system for these.

Emery Radu's conceptual paper applies Hedaa and Törnroos's (2002) two dimensional time–space model, both a chronological approach and kairology – a theory of timing that recognizes performance outcome in its context – to appraise the time management practice of one event manager at two professional basketball matches. Given the pressures on event managers, the research provides a theoretical framework that could provide significant benefits for application in a great many sport event contexts.

The application of the grounded psychological constructs methodology *repertory grid* has been used frequently in the business context, often to educe market knowledge. Here Ensor, Robertson and Ali-Knight have applied the methodology to assess the perception of both established festival Directors in Edinburgh and an expert with responsibility for the strategic development of arts festival at a national level. While the focus has been applied to Edinburgh's portfolio of festivals the constructs elicited from their perception of creativity and innovation for large festivals has the propensity to offer valuable insight for arts festivals elsewhere.

REFERENCES

Getz, D. (2002) Event studies and event management: on becoming and academic discipline, *Journal of Hospitality and Tourism Hospitality*, **9**(1), 12–23.

Hedaa, L. and Törnroos, J. (2002) Towards a theory of timing: kairology in business networks. In R. Whipp, B. Adam, and I. Sabelis (ed) *Making Time and Management in Modern Organizations*, New York, Oxford University Press, pp. 31–45.

Silvers, J.R., Bowdin, G.A.J., O'Toole, W.J. and Nelson, K.B. (2006) Towards an international event management body of knowledge(EMBOK) *Event Management*, **9**, 185–198.

Pine, B.J. and Gilmore, J.H. (1999) *The Experience Economy*. Boston, Harvard Business School.

Woolf, F. (2004) *Partnerships for Learning: A Guide to Evaluating Arts Education Projects*, London, Arts Council of England.

Audience development in the arts: A case study of chamber music

Maxine Barlow and Simon Shibli

INTRODUCTION

Contextual Framework

Arts Council England (ACE) is the national development agency for the arts in England and describes itself as 'bedrock of support' for the arts via the distribution of public money from the Government and the National Lottery.

> Arts Council England works to get more art to more people in more places. We develop and promote the arts across England, acting as an independent body at arm's length from government. (www.artscouncil.org 12 June 2006).

Over the last two decades participation rates in the arts and cultural activities have been static (ONS, 2004). In order to address this issue, ACE recognised the importance of investment in audience development as a device to increase participation rates. An example of ACE's commitment to audience development was the £20 million New Audiences Programme (NAP), which ran between 1998 and 2003, funded by the Department for Culture, Media and Sport. The NAP aimed to bring new audiences to the arts and to take art to new audiences. The programme provided a fertile testing ground for innovative approaches and ideas to help ACE meet its audience development commitment.

This aim to increase the number of people taking part in arts activities is now firmly embedded within Government policy in the UK. The drive to increase participation is not merely a desire but has become a formal requirement. Targets have been agreed between the Department of Culture Media and Sport (the department which funds ACE) and the Treasury for the period 2005 to 2008 in the form of 'Public Service

Agreements'. In the *2005–2008 Public Service Agreement*[1], Public Service Agreement 3 (PSA 3) is defined as

- PSA 3: To increase the take-up of cultural and sporting opportunities by (people aged) 16 and above from priority groups by 2008.

The PSA 3 target will be measured by six indicators. They are

- Increasing the number who participate in active sports at least twelve times a year by 3%, and increasing the number who engage in at least 30 minutes of moderate intensity level sports, at least three times a week by 3%.
- Increasing the number who participate in arts activity at least twice a year by 2% and increasing the number who attends arts events at least twice a year by 3%.
- Increasing the number accessing museums and galleries collections by 2%.
- Increasing the number visiting designated historic environment sites by 3%.

Of particular note for this research is the target to increase the number of people aged 16 and above who attend two or more different arts events at least twice a year by 3%. The baseline figure for the arts participation target is currently being established via the DCMS commissioned survey *Taking Part*[2]. At the time of writing, provisional results (ONS, 2006) from the first six months of a year long survey indicate that the baseline figure for all adults meeting the PSA 3 threshold of attending two or more different types of arts events during the past twelve months was 32.8%. It is therefore clear that attendance at arts events generally (as defined in PSA 3) is a minority activity.

ACE's current corporate plan has reiterated its own and the Government's commitment to increase the number of people who engage with the arts. ACE will invest £1.1 billion of public money from the Government and the National Lottery in supporting the arts, between 2006 and 2008.

It is the view of ACE that the National Lottery has:

...transformed the landscape for audiences and artists, by supporting thousands of projects, both large and small... audiences around the country enjoy new and refurbished arts buildings, and a huge range of arts activity. Arts Council website (2006).

However despite this unprecedented level of investment, participation in the arts has remained relatively static over the past 20 years. The best that can be said about the additional funding for the arts since the inception of the National Lottery in 1994 is that it may have averted a decline in participation rates.

In short, there is a clear commitment to increasing attendance in the arts from those responsible for the creation and implementation of cultural policies within the UK. This paper evaluates a programme of audience development in classical chamber music in order to identify the characteristics of effective audience development techniques.

Classical Music Attendance

Classical music is a broad and somewhat imprecise term, referring to music produced in, or rooted in the traditions of, European art, ecclesiastical and concert music. What the public generally calls 'classical' music is actually many different styles of music that come from many historical periods, however it is generally accepted that classical music encompasses the broad period from roughly 1000 to the present day. The term 'classical music' appeared in the early nineteenth century, and was an attempt to canonize the period from Bach to Beethoven[3]. The modern conception of 'chamber music' may be said to date from Haydn. Chamber music is a form of classical music, written for a small group of instruments which traditionally could be accommodated in a standard room or 'chamber', as opposed to a church or larger building. In broad terms, chamber music encompasses:

> ...any 'art music' that is performed by a small ensemble of instruments with one performer to each part... The word "chamber" signifies that the music can be performed in a small room, often with an intimate atmosphere[4].

An overview of classical music attendance in Great Britain is provided below. It is important to note that because chamber music is but a sub-discipline of the classical music genre, participation rates in chamber music will be lower than the overall participation rate for classical music. Probably as a result of chamber music being a minority activity within 'classical music', there is no national research relating specifically to chamber music participation rates within the public domain.

The data below has been compiled by the Office for National Statistics (2004), and provides a summary of classical music attendance over the past two decades. Between 1986 and 2003 the adult participation rate in classical music has been reported at between 11% and 13%. Any variation within this narrow banding is a function of rounding and sampling error rather than 'real' changes in the demand for attending classical music events.

Research published by ACE (2004) also verified the static level of participation in classical music whereby it was found that 10% of respondents had attended a classical music event in the past 12 months, in both 2001 and 2003. The percentage of respondents who had attended a classical music event in the four weeks prior to the research was 3% in 2001 and 2% in 2003. A further research report focusing on cultural diversity published by the Office for National Statistics (2004), found that 9% of respondents had been to a classical music concert in 2003.

The 2000 Time Use Survey measured the amount of time spent by the UK population on various activities, including leisure activities, through the use of questionnaires and daily diaries. This survey included a specific question on classical music attendance. The findings show that 3.5% of respondents had attended 'a concert or performance of classical music of any kind' in the past four weeks, which is highly comparable with the ACE research of 2001 and 2003 (Arts Council of England, 2004).

Overall, the research findings discussed are relatively consistent: 12 month participation rates are all between 9% and 13%; and four weekly participation rates are between 2% and 3.5%. Aside from these data sources, there appears to be limited information which specifically relates to classical music attendance in England. No published studies focusing specifically on chamber music were identified within the public domain, although there was one unpublished thesis commissioned by Music in the Round (MitR) (Epton, 1998), which provided some demographic data which is used as a basis for comparison with the current research in the results' section.

The research published by ACE in 2004 (Epton, 1998) also highlighted that classical music was characterised by a high level of repeat attendance. The study found that over one third (34%) of attenders had

Table 1 Classical Music Attendance

	1986/87	1996/97	1998/99	2000/01	2001/02	2002/03
Classical Music	12%	12%	11%	12%	12%	13%

Source: Office for National Statistics (2004).

been to a classical music concert three or
more times in the past year (2003). A
rational approach to this sort of finding is
that managers should focus their efforts on
encouraging existing attenders to participate more frequently (market penetration)
rather than persuading non-attenders
to become first time attenders (market
development).

	Existing Products	New Products
Existing Markets	Market Penetration	Product Development
New Markets	Market Development	Diversification

Source: Ansoff, 1965.

Fig. 1. The Ansoff Matrix

Audience Development

To put the current research into context, it is
useful to consider what is meant by the term
'audience development'. The ACE defines
audience development as:

> Activity which is undertaken specifically
> to meet the needs of existing and potential
> audiences and to help arts organisations
> to develop ongoing relationships with
> audiences.[5]

Audience development in the arts is consistent with generic principles of marketing
notably the Ansoff Matrix[6] via its recognition
of the need to serve existing audiences as
well as increasing the participation base.
Developing audiences for the arts embraces
a wide range of potential outcomes. These
outcomes range from increasing the frequency of participation of existing audiences, to broadening the range of people
who attend and participate via the attraction
of first time attenders. An audience development strategy can be aimed at both new and
existing audiences (as will be shown in
figure 1). In terms of existing audiences, strategic approaches can include encouraging
attendance at unfamiliar art forms or event
types, trying new or different venues, attending a venue more frequently or attracting
back-lapsed attenders. Previous research
(Clayton, 1996) found that for infrequent
arts attenders 'choice, value and enjoyment'
were more important than price in motivat-
ing attenders. By understanding and
knowing existing and potential audiences, it
is possible to develop a relationship and

communicate effectively. The challenge of
getting to know potential new audiences is
encapsulated in the two quotes below:

> Many non-attenders of arts events have no
> understanding of, or familiarity with the
> arts. They fear that they will not understand,
> feel overawed, unintelligent and inferior and
> have no reason to believe that they are
> going to enjoy themselves.[7]

> The issue is paying for an unknown quality.
> For example if you go for a pizza you get
> pizza; if you go to the pub you get drunk; if
> you go to the theatre you don't know what
> you are getting. We need to reduce the
> 'unknown factor' rather than the price.[8]

Audience development is a planned process
which involves building a relationship
between an individual and the arts. Research
published by ACE (Hassan, 2004) describes
the fundamental nature of attracting new
audiences:

> It is about giving people an experience that
> inspires, moves or challenges them. It is
> about giving them something they did not
> have before and, more importantly, it is
> about turning a single encounter into a
> long-term affair.

Audience development can be conceptual-
ised using the Ansoff Matrix, a technique normally used to evaluate business strategies,
notably in marketing (Figure 2).

A market penetration strategy relates to
achieving growth with existing products in
their current market segments, aiming to

Fig. 2. The Ansoff Matrix applied to audience development strategies

increase market share. This is the least risky strategic approach since it leverages existing resources and capabilities. Market development seeks growth by targeting existing products to new market segments. Appealing to a new market typically has a higher degree of risk than a market penetration strategy. It is often held that it costs four to five times as much to win one new customer as it does to satisfy an existing customer. If this 'rule of thumb' is true, then it is perhaps not surprising that most marketing effort in the arts seems to be focused on market penetration rather than market development. Developing products targeted to existing market segments is termed product development, again this approach carries more risk than simply attempting to increase market share. Diversification is the most risky of the four growth strategies as it incorporates both product and market development, by developing new products for new markets. This high-risk approach requires the strong chance of a high rate of return and is rarely used in arts marketing. The primary focus of this research is on market penetration and market development.

The Ansoff Matrix provides a logical framework for evaluating strategies designed to impact on customers (audience) and products (programme) in the arts. It provides a conceptual framework for analysing potential audience development strategies, and facilitates decisions regarding the types of strategies to pursue in order to achieve predetermined outcomes.

Different approaches within the Ansoff Matrix require different strategies. As noted previously, it is important to get to know and understand existing and potential audiences, in order to identify the right approaches. In the case of market development (attracting new audiences for the first time) free or low-cost 'taster' sessions and the provision of information and reassurance may be appropriate strategies. Alternatively, the pursuit of a market penetration strategy (increasing attendance frequency and bringing back lapsed attenders), may require different strategic approaches, for example incentivising attendance at a variety of classical music concerts via a subscription offer.

RATIONALE

In 2002, ACE acknowledged the lack of quality chamber music provision in England, by inviting organisations to tender for a National Touring Programme (NTP) grant. The NTP grant would part-fund a project focused on chamber music delivery to address this shortfall in provision.

The National Touring Project used National Lottery funds to support the distribution of work from a broad range of arts disciplines to audiences in England. The focus of the NTP was to encourage dynamic relationships between artists, producers, venues, promoters and audiences. The NTP aimed to encourage new networks, commission new works and explore new ways of presenting work from a diversity of sources and cultural backgrounds. The purpose of the 'Chamber Music Touring Project' was to increase the provision of high-quality chamber music and to establish a strong national brand through which relationships between artists, venues, promoters and audiences could be developed.

MitR was successful in securing NTP funding to manage the Chamber Music Touring Project, and named its project Around the Country (AtC). As the UK's foremost promoter of chamber music outside London (promoting around 150 concerts a year and running a varied range of education projects), MitR has operated from its home at the Crucible Theatre in Sheffield since 1984. The promotion of chamber music in different venues around the UK had long been an ambition of MitR. However, lack of opportunity and funding meant that the ambition remained an aspiration rather than a reality.

At the heart of MitR's ethos is the vision of making chamber music accessible to all – to 'unstuff' chamber music. At MitR concerts, artists do not wear formal dress and are invited to speak to the audiences about the music they play, thereby adding value to the experience of attending a concert and breaking down potential barriers. This contrasts with mainstream classical music, where formal dress and no verbal communication with the audiences are commonplace. The spoken introductions are informal, often witty and always illuminating both for the musically educated and the new or occasional concert goer. The 'in the round' concept which entails the audience sitting 'around' the performers, creates a friendly intimacy and the sense of a shared experience. This effect is summarised by Peter Cropper, Artistic Director of MitR as follows:

> "audiences genuinely feel part of the event – as no one is more than 20 feet away from the performers so much so that you can almost read their music and feel the rosin fly from the bow".[9]

The unique atmosphere of chamber music being played 'in the round' has been a defining factor in the success of MitR in Sheffield since 1984.

The inaugural AtC tour started in Autumn 2003 and featured eight tours to 'in the round' venues, involving eight ensembles and 50 concerts. The tours offered a balance of established and emerging artists from the UK and abroad. The second year of the AtC tour added a further three venues and the presentation of 52 concerts. MitR has worked with the same venues each year on the AtC tour in order to develop continuing relationships with the regional audiences. This network of venues has gradually expanded from 8 in 2003/04, to 14 in 2006/07. It is a measure of the considerable success of the tour that ACE has included AtC in MitR's core funding such that AtC is now an integral part of MitR's activities rather than a project funded on an *ad hoc* basis.

METHODOLOGY

The research was conducted over a two-year period using a survey questionnaire that was distributed at a total of 65 concerts (48 concerts in year one and 17 concerts in year two).

On Music in the Round's inaugural AtC tour in 2003/04, research was carried out at 48 of the 50 concerts. Some 2697 questionnaire packs were distributed at the 48 concerts, and 1417 usable surveys were returned, giving a year one response rate of 53%. During the second year of the AtC tour (2004/05), the evaluation took a different approach, and focused specifically on the three new venues included in the tour. Audience surveys were distributed at the 17 concerts, which took place at the three new venues only. In total 879 questionnaire packs were distributed and from this 431 were returned, giving a year two response rate of 49%.

In terms of the survey distribution, front of house staff at each venue invited randomly selected members of the audience to take a questionnaire pack with them for completion at home, once they had time to reflect on the concert they had just attended. The questionnaire pack contained a covering letter from the Leader of the Lindsays and Artistic Director of MitR, the questionnaire, and a pre-paid business reply envelope for completed surveys to be sent to the independent evaluators.

The covering letter emphasised the independence of the survey and provided assurance that all responses would be anonymous and that any personal details would not be disclosed to third parties. A telephone hot line to the research manager was provided in the event of respondents having any queries about the survey. Finally, an incentive to complete the survey was offered in the form of a free prize draw to win a selection of signed CDs from some of the ensembles that took part in the tour (Figure 3).

The number of responses received from each venue is in the first instance a function of the number of concerts (which varied from three to seven), the number of questionnaire packs distributed at each concert (due to the principle of diminishing returns less packs may have been handed out as the tour progressed because repeat attenders would have already completed a questionnaire), and the inclination of respondents to complete the survey. The overall response rate of 51% (1848 returns from 3576 questionnaires distributed) is a good response rate from a postal survey conducted in an interested population with no-follow up campaign.

The data from the 1417 questionnaires from year one and the 431 questionnaires from year two was combined to form one SPSS dataset (providing a total of 1848 respondents). The results and discussion which follow are based on the data generated from these 1848 responses.

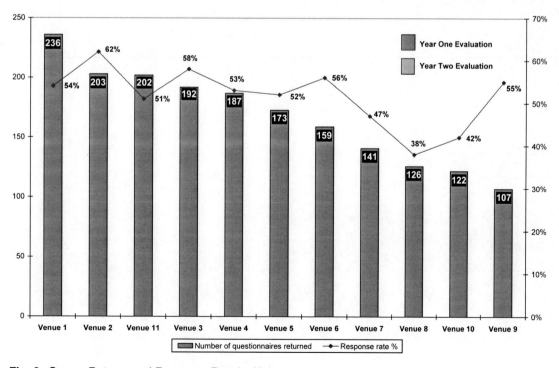

Fig. 3. Survey Returns and Response Rate by Value

The quantitative research was also supported by interviews with key stakeholders within each of the venues and also with MitR staff.

RESULTS

An overview of the demographic profile of the audiences is provided in Figure 4.

It is useful to begin this analysis by considering the demographic profile of the MitR audience. Overall, the audience profile confirms what is known about classical music attenders generally from national level research (ACE, 2002) and chamber music attenders specifically from local level research in Sheffield (Epton, 1998).

In essence, chamber music audiences tend to be older than mainstream classical music audiences. Research by Sheffield University (1997, cited in Epton 1998) and Epton (1998) found that over 80% of chamber music audiences were aged 45 and

over (85% and 81%, respectively). All measures of central tendency suggest that the average age of AtC respondents was 62 (Epton 1998 also 62) and the range of ages found in the current survey was from 16 to 93. It was estimated that at least 44% of respondents were retired, which appears to reflect the age profile shown above and is comparable with the findings from Sheffield University's research (41%). Chamber music audiences also tend to have greater concentrations of people drawn from socio-economic groups AB than classical music audiences. The current research findings highlighted that the majority (78%) of the MitR audience were drawn from socio-economic groups AB, which is comparable to Epton (85%).

In the second year of this evaluation, an additional question was added to the survey questionnaire to measure the frequency of attendance. On average, respondents had attended, or planned to attend, more than three concerts during the

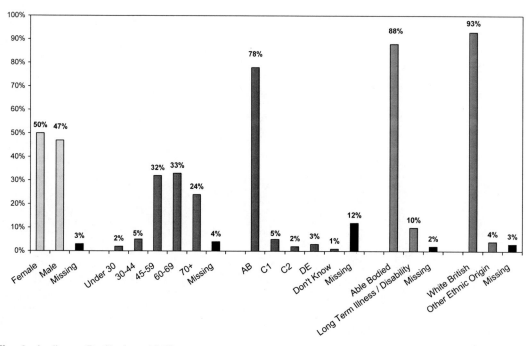

Fig. 4. Audience Profile (*n* = 1848)

2004/05 AtC series, at the three new venues added to the tour. The findings concurred with those of ACE (2004), which highlighted that classical music was characterised by a relatively high level of repeat attendance.

One of the goals of the AtC tour was to have an 'audience development' effect, where audience development is defined by 'market development' that is, to broaden the base of people who attend chamber music concerts. The research evaluation measured the success of this development aim to attract first time attendees. In total, 3.7% of the respondents were 'new to genre', i.e. they had never attended a chamber music concert before. This equated to 68 individuals from the sample of 1848. In the current climate of static participation rates, this is a noteworthy achievement. However, it should be noted that first time attenders at chamber music concerts does not also mean first time attenders at a classical music concert. It is quite possible, although unproven, that the 'new to genre' respondents had attended a mainstream classical music concert previously but were first time attenders at a chamber music concert.

Across the full sample of 1848 respondents, an average of 3.7% were first time attenders at a chamber music concert and 13.9% were visiting the venue at which they received the survey pack for the first time. Table 2 summarises the market development data by venue.

The market development data is now subject to further analysis to evaluate the quality of this effect through analysis of the specific marketing strategies pursued. Figure 5 begins this process of further analysis by illustrating how the 'new to genre' effect varied by venue.

Figure 5 highlights that there is significant variation in the 'new to genre' effect experienced at the 11 different venues. The horizontal line represents the average score (3.7%).

Three venues (Venue 3 = 0%, Venue 5 = 1.7% and Venue 6 = 1.9%) have well

Table 2 Audience Development Effects

Venue	New to genre (%)	New to venue (%)	No. of respondents
Venue 1	3.5	7.9	236
Venue 2	2.6	11.8	203
Venue 3	0.0	11.0	192
Venue 4	6.4	4.3	187
Venue 5	1.7	24.7	173
Venue 6	1.9	26.9	159
Venue 7	4.3	5.0	141
Venue 8	5.6	10.6	126
Venue 9	2.9	1.0	107
Venue 10	5.0	39.5	122
Venue 11	7.7	14.9	202
Overall	3.7	13.9	1848

below average levels of first time attenders. These findings are perhaps not surprising as the venues concerned all had strong local chamber music provision and their marketing effort was focused primarily at existing customers, that is, a market penetration strategy.

In the locality of Venue 3 there was a strong existing chamber music audience however the existing network of provision was becoming financially unsustainable. The launch of the AtC series at this venue met the needs of the pre-existing audience but no serious efforts were made to attract first time attenders.

The specific local area in which Venue 6 was situated had no prior chamber music provision however the neighbouring area has a very strong tradition of classical and chamber music. It would appear that this strong audience base was persuaded to travel to a new venue in the neighbouring town by the appeal of the quality of this provision. Likewise, Venue 5 had a strong pre-existing audience base that was attracted to attend the AtC series at a church venue without a history of hosting chamber music concerts. At both Venues 5 and 6 the aim of the local management was to bring an existing audience to a new

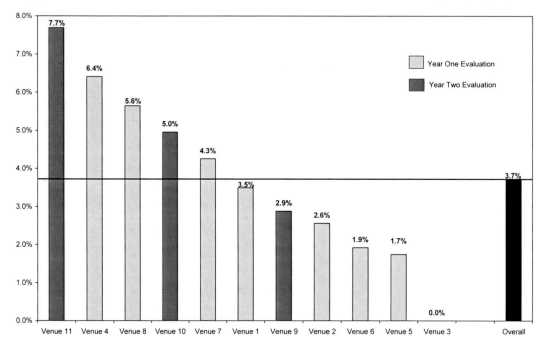

Fig. 5. Percentage of First Time Attenders 'new to venue'.

venue (product development) rather than to attract a new audience (market development). The relatively low 'new to genre' score at Venue 2 was in part attributable to the remote geographical location of the venue. The venue locality was described by one stakeholder as 'quite insular' with the core audience made up of a tight group of 'friends'. For this reason a market development strategy was not pursued. The travelling distance from other local towns was described as 'prohibitive' in the sense that local people in the vicinity tended to travel only small distances to attend events, or chose to travel further afield to take a 'day trip', which would incorporate attendance at a concert. The journey time for much of the potential catchment area fell outside what would be considered a reasonable commute to attend an evening concert, but not far enough away to consider taking a longer trip to include the concert.

By contrast, Venue 4 (6.4%), Venue 8 (5.6%), Venue 10 (5.0%) and Venue 11 (7.7%) all had

significantly higher levels of first time attenders than the average (3.7%). Possible reasons contributing to this successful audience development are discussed below.

Investigation of local factors indicated that the high level of new attenders at Venue 4 was likely to be attributable to audience cross-over from other art forms and the lack of provision for chamber music audiences locally (e.g. Venue 4 was a theatre with a strong drama programme but only a limited music programme). Further investigation of the marketing techniques utilised by Venue 4 identified that the venue's publicity materials were very clear and concise and that all events were given equal prominence within the generic venue brochure, which was viewed positively by MitR. This type of effect is often termed 'cross-fertilisation' of audiences and recognises that attenders of certain art forms are more likely to attend other art forms than the population as a whole. For example, there is a high-audience cross-over between classical music and opera. In addition, MitR reported that within the generic brochures produced at several other venues, the AtC series of concerts had been clustered together at the rear of brochures, rather than being integrated into the main body of the brochures. This positioning was perceived to be less effective, as potential audiences looking for events in chronological order may miss forthcoming AtC concerts and may also have had a marginalising effect. Venue 4 also took an innovative approach to the production of AtC publicity material and utilised additional marketing techniques such as poster campaigns.

The promotion of a series of chamber music concerts was a relatively new phenomenon at Venues 10 and 11. Venue 10 was successful both in attracting first time attenders and attracting the existing chamber music audience in the locality to try a new venue (through networking with the local chamber music society). The new venue was not a 'typical' chamber music venue in the traditional sense, as it was situated on a school site. It is likely that the increased accessibility, resulting from using a less formal venue that would not intimidate or overawe newcomers contributed to the venue's success in terms of new audiences. There is also evidence of audience cross-over specifically at Venue 11, where the only other musical activity in the area is provided by a local Symphony Orchestra. Prior to the launch of the AtC series, chamber music audiences in this area had been forced to travel over 30 miles to the next large city or over 70 miles to London in order to attend chamber music concerts. Further to these factors, Venue 11 was also an open public venue which may have been perceived as less intimidating for potential new audiences. The environment in this locality was also particularly amenable to marketing via 'word of mouth'. A change in marketing strategy from targeting existing subscribers (market penetration) to a wider strategy directed at the general public

(market development) is also likely to have been an influential audience development factor at these venues.

MitR reported that from its experience of delivering the AtC series, the key factors which contribute to successful audience development are: knowing your audience, being prepared to try new techniques and commitment.

Table 3 categorises the 'new to genre' scores into 'high', 'average' and 'low' effects and suggests possible factors and strategic marketing approaches that may have contributed to these scores.

In the same way that the proportion of first time attenders varied by venue so too they varied by age category, as shown in Figure 6.

People aged 'under 30' formed 2% of the sample yet 22% of the first time attenders were found in this age group. By contrast, people aged '60–69' formed 35% of the audience and only 2% of these respondents were first time attenders. In short, there was an inverse relationship between age category and the incidence of first time attendance at a chamber music concert. This provided an insight into the quality of the audience development effect in terms of it being particularly pronounced in the relatively younger age categories. It is worth noting that although young people (aged under 16) did attend some of the concerts, it was agreed not to interview minors. Therefore, the overall audience development effect may actually be understated. However, qualitative comments were made by customers who noted the apparent audience development effect with young people, such as:

> I was pleased that many members of the audience were young – unlike my recent experiences of chamber concerts. This must be good for the future.

To gauge a complete picture of market development, the proportion of first time attendance at venues (new to venue) was also measured. The sample average was 13.9%.

Table 3 New to genre summary

'New to genre' score	Venue	Contributory factors to new to genre score
High 'new to genre' score **(5.0% – 7.7%)**	Venue 4 Venue 8 Venue 10 Venue 11	Audience cross-over New phenomenon Low existing provisions Awareness campaigns/innovative advertising Targeting the general public (non members) Strong commitment of key stakeholders
Average 'new to genre' score **(2.9% – 4.3%)**	Venue 1 Venue 7 Venue 9	Some targeting of the general public Moderate existing chamber music provision Promotion through existing channels
Low 'new to genre' score **(0.0% – 2.6%)**	Venue 2 Venue 3 Venue 5 Venue 6	Targeting existing customers/subscribers Strong local chamber music provision Predominantly subscription/society members Promotion through existing mailing lists

This means that for almost one in seven respondents, it was their first visit to the venue concerned. In the same way that the 'new to genre' data was not evenly distributed across venues, so too the 'new to venue' data is distributed unevenly as shown in Figure 7.

The 'new to venue' statistic of 13.9% for the full sample is driven by the responses from Venue 5 (24.7%), Venue 6 (26.9%) and Venue 10 (39.5%). Analysis of local factors specific to Venues 5 and 6 revealed that they both pursued a successful product

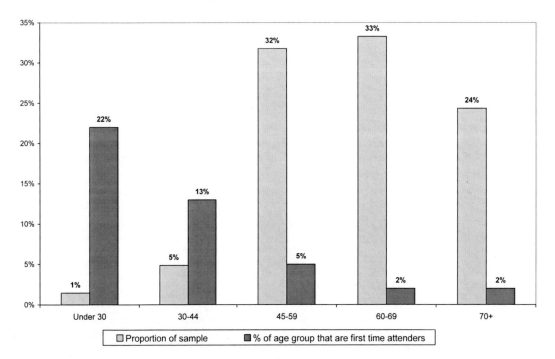

Fig. 6. First Time Attenders Relative to Age Categories

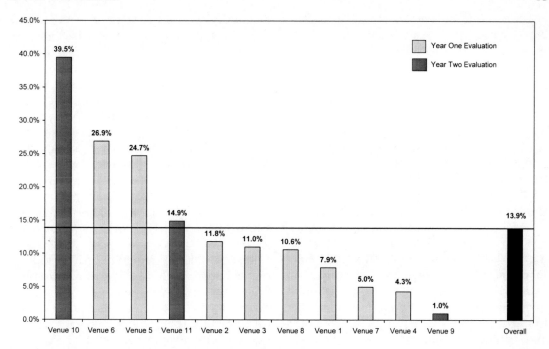

Fig. 7. Percentage of First Time Attenders 'new to venue'

development strategy. Venue 5 is not a full time professional venue and is based within a school. Prior to AtC, both venues were used infrequently to host chamber music concerts. Therefore, a major objective for the venues was to build an audience at venues that were relatively new to chamber music provision. There were other venues in both localities that had a strong reputation in this respect. Thus it would appear that the existing chamber music audiences were successfully persuaded to try out a new venue (product development). In contrast to these 'new to venue' achievements, both venues 5 and 6 experienced low 'new to genre' scores, indicating that market development strategies were either not prioritised, or proved to be ineffective.

Venue 10 was on the site of a private school that had shown enthusiasm for greater integration with the local community. It would appear that the venue has been successful with this aim, with almost

four out of every ten respondents being a first time visitor. The qualitative feedback relating to this new venue was also highly positive:

Delightful small auditorium–looking forward to attending more here next season.

Good venue, appropriate size, great acoustics and a vibrant, absorbing performance. In all, a thoroughly enjoyable event.

In contrast to the above findings, venues 4 and 7 both generated high 'new to genre' scores (6.4% and 4.3%, respectively) but experienced low 'new to venue' effects (4.3% and 5.0%, respectively). These findings suggest that the venues concerned have been particularly successful in generating first time attenders to chamber music (market development) and relatively unsuccessful in attracting customers to new venues (product development). It is likely that the cause of these findings is attributable to successful strategic marketing resulting in cross-fertilisation of audiences from other art forms as there was no existing chamber music provision and hence no formally identified audience in the area.

Further investigation of this 'new to genre' audience development effect is undertaken by analysing how first time attenders first found out about the concerts, relative to those who had attended a chamber music concert previously. Bearing in mind that the sub-sample of first time attenders is 68 out of a total of 1848, the data for this analysis is presented in Table 4. To aid interpretation, Table 4 collapses the six main marketing devices used into three generic categories, namely: direct mail, distribution and other means. The index figure computes the relationship between first time attenders and 'other' customers (those that have attended a chamber music concert previously) for each category. An index figure of 100 would reflect equal proportions for first time attenders and others, whereas an index score below 100 reflects an under-representation of first time attenders and an index score over 100 reflects an over-representation.

Table 4 shows that word of mouth (somebody told you about it), and 'other' are the means by which first time attenders were most likely to find out about the concerts. By contrast, people who attended a chamber music concert previously were most likely to have found out about the concerts by direct mail. Word of mouth is perceived by MitR as one of the most important marketing tools because it is developed by building trust. It is important to the MitR brand that effective 'word of mouth' marketing is taking place in relation to the AtC concerts. Overall 24% of first time attenders found out about the concert via word of mouth, in comparison to 13% of 'other' attenders.

In terms of marketing via 'distribution', a higher proportion of first time attenders found out about the concerts via a generic marketing brochure picked up at a venue, rather than through the AtC-specific marketing brochure. By contrast, existing attenders were more like to have found out about the concert from the specific AtC brochure, rather than generic venue marketing. This emphasises the potential for encouraging audience cross-over by promoting AtC concerts through generic marketing tools where possible. Encouraging audience cross-over is likely to be more effective than targeting the general public as a whole, as those customers already have

Table 4 How attenders first found out about concerts

Marketing tool	First time attenders	Others	Index[1]
ATC brochure sent by post	12%	16%	75
Venue brochure sent by post	13%	27%	49
Direct Mail	**25%**	**43%**	58
ATC brochure picked up locally	9%	13%	69
Venue brochure picked up locally	16%	10%	166
Distribution	**25%**	**23%**	111
Somebody told you about it	24%	13%	174
Other	26%	21%	125
Other (not mail / distribution)	**50%**	**35%**	144

[1]Index = ((First time Attenders / Others) * 100))

some form of engagement with the performing arts (depending on which venue they attended this could include drama, music, theatre, dance or comedy) and are therefore less likely to be indifferent or hostile towards the performing arts as a whole.

The marketing campaigns at each of the 11 different venues were venue-specific rather than standardised. Therefore, it would be reasonable to expect that the way in which people first found out about concerts may vary by venue. This proved to be the case as shown in Figure 8.

This data again highlights significant variation by venue, in terms of marketing effectiveness. It is evident that the majority of venues which experienced high 'new to genre' ratings (particularly Venue 8, Venue 10 and Venue 11) also had high proportions of audiences finding out about the concerts via 'other' means. Promotion by word of mouth was a key component of 'other means'. Further investigation highlighted

that almost one quarter of the audience at both Venue 10 and Venue 11 (25% and 24%, respectively) had first found out about the AtC concerts via word of mouth. MitR confirmed that through their personal experience and local knowledge of these venues, advertising via word of mouth and building relationships with audiences via the 'personal touch' was likely to have been a strong contributing factor to the successful audience development at these venues.

The above points are further evidence to suggest that there is no 'one size fits all' standardised approach that can be implemented to generate successful audience development. Successful marketing strategies must be specific to the local environment and work to the strengths of the existing infrastructures.

Table 5 brings together all these variable factors and provides a profile of the marketing effectiveness and market development effects for each venue.

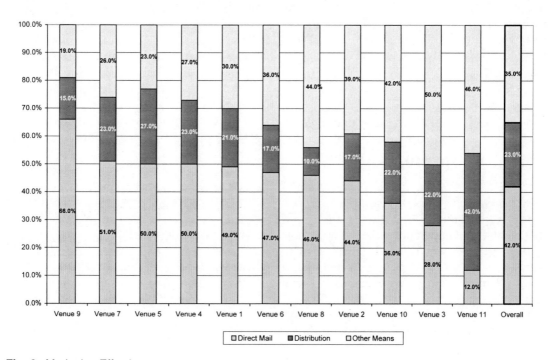

Fig. 8. Marketing Effectiveness

Table 5 : Marketing profiles by venue

VENUE	MARKET DEVELOPMENT		AGE (proportion aged under 45)	MARKETING		
	NEW TO GENRE	NEW TO VENUE		DIRECT MAIL	DISTRIBUTION	'OTHER'
Venue 10	High	High	High	Low	Average	High
Venue 11	High	Average	Average	Low	High	High
Venue 8	High	Average	High	Average	Low	High
Venue 4	High	Low	Average	Average	Average	Low
Venue 1	Average	Low	Average	Average	Average	Average
Venue 7	Average	Low	Average	Average	Average	Low
Venue 9	Average	Low	Low	High	Low	Low
Venue 5	Low	High	Low	Average	High	Low
Venue 6	Low	High	High	Average	Low	Average
Venue 2	Low	Average	High	Average	Low	Average
Venue 3	Low	Average	Average	Low	Average	High
Overall	**3.7%**	**13.9%**	**62** (average age)	**42%**	**23%**	**35%**

Venues scoring highly on market development in terms of 'new to genre' tend to have low / average levels of success with direct marketing, and higher levels of success through 'other' marketing techniques, a strong element of which is marketing via word of mouth. Overall, this table suggests that a pre-determined formula that can be adopted to guarantee successful market development does not exist.

One striking feature of audience development is that one size does not fit all when it comes to the type of audience development organisations could and should undertake.[10]

SUMMARY

The most successful venue in terms of audience development is arguably Venue 10 (Table 6). Around 5% of the audience at this venue had never attended a chamber music concert before and a further 40% were new visitors to the venue. As discussed in the results' section, this exceptional 'new to venue' score may be related to the fact that this venue was considered to be accessible, informal, friendly and welcoming, both to previous chamber music attenders and new audiences alike. The successful

formula for audience development at this venue was achieved through integrating a range of factors that were appropriate for the local context, notably: networking with local arts organisations; building on the infrastructure created by the existing (but limited) chamber music provision; and, building support via word of mouth. By contrast, Venue 11 (7.7% new to genre/14.9% new to venue) was located in an area with no existing chamber music provision and therefore this venue started a brand new programme which in theory takes a high-risk strategic approach involving market development and diversification. In this

Table 6 Audience Development Summary

		New to venue		
		Low	Average	High
New to venue	**High**	4	11	10
			8	
	Average	1		
		7		
		9		
	Low		2	5
			3	6

case, a high level of commitment and enthusiasm for engaging with new audiences and a willingness to try something new, were the elements which are believed to have contributed to successful audience development.

CONCLUSIONS

Music in the Round's AtC series in both 2003 and 2004 achieved a significant and positive audience development effect in terms of attracting first time attenders to a chamber music concert. Overall 3.7% (68 people) of a sample of 1848 respondents were making their first attendance at a chamber music concert. When considered alongside the national context of participation rates that have been static for 20 years, this is a notable achievement.

In terms of marketing effectiveness, there were clear differences in the way that first time attenders and existing chamber music audiences found out about the AtC series. Word of mouth and 'other' are the means by which first time attenders were most likely to find out about the concerts. By contrast, people who attended a chamber music concert previously were most likely to have found out about the concerts by direct mail. Market development requires managers to adopt a more innovative approach to marketing and is unlikely to be achieved via the conventional mode of direct mail or venue-based distribution channels. To reach new customers requires going beyond the comfort zone of doing what has always been done in the past.

Whilst it is recognised that there cannot be a standardised approach to market development, due to the influence of local demographics, existing provision, perceived accessibility and the strategic priorities of the venue, there are several contributory factors that can be outlined as good practice in market development.

A strategic approach of pursuing audience cross-over has proved successful.

Encouraging audience cross-over is likely to be more effective than targeting the general public as a whole, as those customers already have some form of engagement with the performing arts and are therefore less likely to be indifferent or hostile towards the performing arts. It would appear that the route to succeeding in this approach stems from integrating new products within generic publicity materials, and further marketing via distribution through new channels and building up trust for the product via word of mouth. At a macro level there are some lessons of good practice that can be rolled out more widely as ACE sets about achieving its contractual targets under PSA 3. In terms of achieving PSA 3, which is as much about the range of activities in which people take part as the absolute number of participants, cross-fertilisation of audiences should prove to be an effective strategy.

An overview of the strategic approach adopted by each venue is summarised below in Figure 9. It is encouraging to note that many venues operated outside the comfort zones offered by the lower risk strategies (market penetration and product development) and successfully pursued market development approaches.

Successful audience development requires a strong commitment to working beyond the usual channels of conventional arts marketing. A willingness to try new approaches and undertake innovative advertising are also

Market Penetration New programme, increased provision Venue 1, Venue 2, Venue 3, Venue 9	Product Development New venues, new programme Venue 5, Venue 6, Venue 10
Market Development New audience, cross fertilisation, innovative marketing Venue 4, Venue 7, Venue 8, Venue 10, Venue 11	Diversification New programme for first time attenders Venue 11

Fig. 9. Audience Development Strategies Pursued Source: Adapted from Ansoff, 1965.

key factors. Furthermore, during the interviews with venue managers and artistic directors, it was identified that there was a strong need for effective collaborative work between the venues, promoters and MitR. Assistance from MitR, including excellent communication, high-quality marketing materials/joint press work, encouragement, support and the sharing of good ideas, was highly commended by the venues. In mainstream marketing this is comparable to a franchisee buying into an existing brand, rather than trying to start a business from scratch.

On a cautionary note, it is appropriate to put the achievements of the AtC tour into a realistic perspective. The 68 (3.7% of the sample) first time attenders to a chamber music concert were not necessarily first time attenders to a classical music concert. Therefore, the effect that may have been measured was simply existing classical music audiences sampling a niche product within the overall genre of classical music. If this was the case, then an audience development effect of 3.7% puts into stark contrast the effort and resources required to encourage those who currently have no engagement with the arts to become arts consumers.

ENDNOTES

1. DCMS (2005) 2005–2008 Public Service Agreement: Technical Note, DCMS, London.
2. Office for National Statistics (2006) *Taking Part: The National Survey of Culture, Leisure & Sport*, provisional results from the first six months of the 2005/2006 survey, ONS, London.
3. http://www.classicalarchives.com/dict/chamber_music.htm accessed 26, June 2006.
4. http://www.answers.com/topic/chamber-music – accessed 26 June 2006.
5. Arts Council England (2005) *Audience development and marketing – Grants for the arts*, http://www.artscouncil.org.uk.
6. Ansoff, H. Igor. (1965) *Corporate Strategy*. New York, McGraw-Hill, p. 109.
7. NOP Market Research (1991) *Report on qualitative research into the public's attitudes to 'the arts'*, Cited in 'A Guide to Audience Development' Maitland (1987). The Arts Council of England.
8. Arts Council of England (2003). *New Audiences: An evaluation of the role and impact of the new audiences programme on participating organisations.* Arts Intelligence, cited in Maitland (1987).
9. www.musicintheround.co.uk.
10. Communication and audience development: A report highlighting communication and promotional tools taken from Arts Council England's New Audiences Programme. Aspirational Arts Partnerships, November 2003. P. 5.

REFERENCES

Arts Council of England (2002) *Arts in England: Attendance, Participation and Attitudes in 2001*, London.

Arts Council of England (2003) *Focus on Cultural Diversity: the Arts in England. Attendance, Participation and Attitudes*. Research Report 34. London.

Arts Council of England. (2004) *Arts in England 2003: attendance, participation and attitudes*. Research Report 37.

Arts Council of England, available at: http://www.artscouncil.org.uk/aboutus/index/php, accessed 12 June 2006.

Arts Council of England. http://www.artscouncil.org.uk/aboutus/investment.php, available at: accessed 12 June 2006.

Clayton, H. (1996) Qualitative research among infrequent attenders at the Crucible Theatre Sheffield and Playhouse Theatre Leeds, Report to Arts Council.

Epton, J. (1998) A study researching the effectiveness of current pricing strategies in satisfying customer perceptions of value and how these perceptions can be changed at Sheffield MitR, MA dissertation, University of Sheffield.

Hassan, A. (2004). *I liked everything – celebrating new audiences*. Arts Council England, North West.

Maitland, H. (1997) *A guide to audience development*, London, Arts Council England.

Office for National Statistics (2004). *Social Trends 34 – Attendance at Cultural Events*, p. 201. Data source: Target Group Index, BMRB International.

'Never Let Me Down Again'[1]: Loyal customer attitudes towards ticket distribution channels for live music events: a netnographic exploration of the US leg of the Depeche Mode 2005–2006 World Tour

Zuleika Beaven and Chantal Laws

INTRODUCTION

I guess money, not devotion, is the only thing here that counts in large amounts
(Quote from Depeche Mode Message Board user N2TheBlu, posted 11:16 am MST, 13 September 2005).

In this paper, we examine contemporary issues in the development of ticketing for live music events, with specific regard to the adoption of new technologies to enable virtual ticket distribution. We are interested particularly in the response of fan communities to these developments, in terms of the impact that electronic distribution may have on brand positioning, perceptions of service quality and fit with loyal customer expectations.

Our approach in establishing a framework for the study has been deliberately and necessarily eclectic, as the academic field of events management is an emergent one (Silvers *et al.*, 2006). Particularly when looking at the consumption of arts events, there are a number of theoretical perspectives that one might adopt, ranging from leisure studies to marketing and consumer behaviour, reception theory, media studies and cultural theory (Roberts, 2004; Leppert and McClary, 1987; Yeoman *et al.*, 2004; Hills, 2002). This is further compounded by the contended nature of fandom itself, and the sometimes disparaging treatment of fandom and fan behaviours in academic literature, of which Hills (2002) and Sandvoss (2005) provide a thorough consideration.

Although we acknowledge the influence of Bourdieu's notion of cultural capital and the dialectic of value set out by Adorno on any consideration of music consumption (as explored by Hills, 2002), the focus here is on the physical process of facilitating and managing a live music event in real-time. Therefore, in the absence of research with concert attenders, we have chosen to draw principally on studies of sports fans, as the levels of engagement by the fan community, and the management processes required are, we suggest, closely congruent.

BACKGROUND TO THE STUDY

During 2005, we conducted a multi-method study, drawing on Kozinets' netnography methodology (2002), to investigate the responses of fans of rock group Depeche Mode to innovative changes in ticketing for the US leg of the band's world tour. The study involved primary analysis of posts to the band's internet message board (IMB).

Rapid developments in technology convergence represent an opportunity for event organisers. The ability to integrate marketing, sales and distribution into single transactions has begun to develop through the use of the internet. In tandem, MP3 standardised and compressed digital file technology, alongside the proliferation of broadband internet connections, have undermined the compact disc, with downloads of music tracks estimated at one billion songs each week in 2004 (Connolly & Krueger, 2005). So, within the rock/pop genre, the shift in music distribution channels has, far from delivering the death of the music industry as initially predicted, begun to offer new and increasingly sophisticated opportunities.

The case we present here was revolutionary in being the first ever use of technology to integrate pre-sale ticketing with download sales of an album. Pre-sale is the process of allocating a portion of tickets to an event for advance purchase by attenders, before they go on general release. While there is generally a cost associated with gaining access to a pre-sale, typically through a membership or fan club fee, it is presented as a service to loyal members wishing to purchase high quality seats. Perception of the quality of the event is linked to the ability of serious fans to access 'good' allocations. Unlike the theatre or classical concerts, where the best seats can be purchased for a high ticket price, the absence of differential pricing at rock concerts means good seats are primarily secured by early purchase (or, later, through secondary and re-sale markets dominated by brokers, touts or scalpers). The speed of internet distribution channels has amplified this process so that for the most popular acts, or for those artists such as Depeche Mode who over 25 years of touring have acquired a very dedicated fan base (Miller, 2003), rapid purchase is considered essential.

So, pre-sales are an established opportunity for event promoters and artists and, as ticket distribution channels have shifted towards online selling, are ripe for technological development. In launching the offer, the band's manager, Jonathan Kessler said:

> Depeche Mode is once again pushing the wire for its fans and utilising cutting edge technology to lead toward a better and faster way for fans to access new music and concert tickets.
> (BBC News, 2005, paragraph 9).

However, as Hills (2002) has argued, whilst niche marketing may have the intention of aligning production and consumption through the adoption of authentic fan values–this can only ever occur in the context of managerial activity that seeks to exploit fan communities for commercial gain. There is, therefore, an inherent tension in an offer designed to simultaneously add value to the fan experience while facilitating improved supply, and it is this polarity that we explore in more detail below.

LITERATURE REVIEW

Fandom, Loyalty and Consumer Culture

In today's culture of mediated information and conspicuous consumption, the intense relationships and extreme behaviour patterns that can typify subcultural or 'fan' behaviours have become the mainstream (Hills, 2002; Ross and Nightingale, 2003; Sandvoss, 2005) . Fascination with a particular pop band, actor or celebrity marks a rite of passage from childhood to adult life. Fandom also provides a space in which people can express their passionate interests in confidence: Star Trek fans who enjoy role-playing as Klingon warriors are an example of behaviour that may appear aberrant or irrational outside of this particular community. Recent research has shown that increasingly communities of interest will be shaped by shared hobbies and leisure choices rather than geographical location, and that increased take-up of the internet is key (Kozinets, 1999). However, the massive increase in media output that typifies an information economy has led people to be more selective in their choices and behaviours. Trends such as polyglotting, authentiseeking, and connoisseurship, as summarised in Table 1, are ideally suited to the internet, where a consumer's identity 'is becoming increasingly complex, actively self-defined and derived from a wider and wider range of influences' (ACE/Henley Centre, 2000, p. 54).

The increased volume and range of entertainment commodities has also necessitated a paradigm shift in marketing and customer relationship management (CRM) strategies as the notion of an experience economy (Pine and Gilmore, 1999) has gained currency. This is due to the increasing importance of entertainment to the global economy and also to the influence of American culture, where there is a tendency for consumers 'to develop a fanaticism towards entertainers and to select consumption as one avenue through which they express this fanaticism'. (Thorne & Bruner, 2006, pp. 51–52). This idea of *consumer culture* has been the focus of much debate, yet the behavioural and attitudinal factors that influence fanatical consumer decisions have received more patchy consideration (Redden and Steiner, 2000).

An alternative theoretical perspective to consumption behaviour is musicological reception theory, which provides a framework for understanding the social and cultural roles that music may assume post-composition, and which, therefore, suggests that fan behaviour will develop a dynamic independent of the original artistic context. According to the analysis of the artist Jeremy Deller (2006), this phenomenon is evident in the behaviour seen in the film he made about Depeche Mode fans in Eastern Europe:

> In Russia, 9 May is Victory Day, a national holiday, but the date coincides with [Depeche Mode lead singer] Dave Gahan's birthday, so it is also known throughout the country as 'Dave Day'. We have some incredible footage of 'Dave Day' in 1992, with thousands of young Muscovites joined in a mass singalong to Depeche songs. To be able to gather and to celebrate their favourite band provided some kind of emotional release after the hardships of communism. Russian teens enjoyed a cultural emancipation with Depeche Mode akin to that experienced by American youth through Elvis and British youth through the Beatles a decade later.

The intense emotional experiences that typify serious fan experiences often appear bizarre and extreme to commentators outside of that fandom; however, they are commonplace and relatively benign

Table 1 Consumer change trends

Trend	Definition	Drivers	Examples
Polyglotting	• Complex and actively self-defined identities drawing on a varied range of sources	• Breakdown in attitudes and social rules e.g. gender, religion, class and education • Media exposure to a more diverse range of influences, lifestyles and cultures • Availability	• Growing retail sales of ethnic food, clothing, toiletries and home decoration • Frequency and extent of foreign travel
Authentiseeking	• Growing desire to obtain the original, the true or the real thing in a mass market world	• Decline in trust in traditional companies and structures • Suspicion of commercial companies and their products • Reassurance of the 'authentic' label • Desire to be individual • Growing experience economy	• Growth in 'real' holidays • Flourishing crafts industry • Diminishing appeal of brands which do not feel authentic
Connoisseurship	• We are all connoisseurs in one way or the other and anything can be the subject of connoisseurship	• Rising income makes it possible to indulge in a passion and still pay for life's necessities • The increased need for self-expression and to construct identity	• Growth in specialist retailers • Strength of word of mouth as a communication channel
Perfect moments	• A moment in our lives when the reality lives up to the fantasy or when an everyday is heightened above the mundane	• Growing wealth means that standards and expectations are being raised	• Marketers and advertisers using perfect moments to sell products

People as players	• Play as a vehicle to define people and allow them to escape from the worries and pressures of everyday life	• People are now seeking and expecting guaranteed perfect moments • Media exposure to celebrity lifestyles • A rise in average disposable income	• Leisure providers safeguarding perfect moments through door policies and targeting • British consumers now spend more on leisure activities than food, rent and rates • Blurring of retail and leisure
Communal yearning	• 'Community' is still an aspiration, but one which is being redefined and becoming more personalised	• The move from needs to wants • A world becoming more mass-produced and uniform • Growth of new 'interactive' media • Increased mobility • New media	• Internet communities • Non-geographic communities (e.g. Harley Davidson club)

Source: Selectively adapted from ACE/Henley Centre, 2000: pp. 54–60.

expressions of fan devotion. Academic treatment of extreme fanatical behaviour within particular communities of interest, sports fans (Hunt *et al.*, 1999) or science-fiction aficionados (Kozinets, 2001) for example, tends to give weight to extremes of behaviour (such as football or soccer hooliganism) that actually typifies only a small subset of attendees. Thorne & Bruner (2006) argue that, as celebrity becomes more embedded in everyday life, the experience of fandom is necessarily a more mainstream cultural and social phenomenon. Furthermore, the terminology itself presents problems, as the label of fanatic has negative connotations. It is, therefore, necessary to consider the definitional vocabulary, and Table 2 presents a mapping exercise undertaken from key authors in the field of sports fan studies to produce a terminology appropriate to rock music fans. Of the four rock music fan typologies we identify, loyal, die hard and some dysfunctional fans were represented within the study, and we adopt 'serious fan' as a collective term for them.

Thorne and Bruner (2006, pp. 53–55) have further identified four characteristics that are common to fans across different segments. They hypothesise that fans can be differentiated by the intensity of their level of fanaticism, expressed through and gauged by these characteristics. One aspect not considered by Thorne and Bruner is the issue of brand consumer loyalty. Hunt *et al.* (1999, p. 259) identify three contributory factors, which are: 1) perceived brand performance fit; 2) social and economic identification with the brand and 3) habit and long history with the brand.

The motivations of fans to engage with a particular object and their journey to a deep engagement can, therefore, be seen to be complex, emotional and irrational in many ways (Hills, 2002; Sandvoss, 2005). The bond of trust that fans form with their object of fascination would appear to be the key in sustaining a lasting service relationship, and research into audiences for the arts (Hayes and Slater, 2002) has indicated that nurturing loyalty in the long-term should be a key strategic concern and essential for competitive advantage. Discussion of brand trust (Delgado-Ballesster and Munuera-Alemán, 2001) can be regarded as an issue of perceived quality, satisfaction and loyalty – it might also focus on the hedonic-emotive aspects of the service journey or attempt to capture the talismatic relationship formed between a fan and the object. Trust has been identified as a key factor in consumer commitment, tolerance for change (such as price increases) and crucial for successful customer–provider encounters or 'moments-of-truth' (Gyimóthy, 2000).

The challenges that highly individualised and irrational consumer behaviours/attitudes present to service providers are that service perception is likely to be more holistic, experiential and hyper-real. Thus, a conventional supply-based delivery structure may present greater potential for negative critical incidents. In the music industry established distribution channels where, 'artists create music, record labels promote and distribute it and the fans consume it' (Graham *et al.*, 2004, p. 1087) are being challenged by the burgeoning of virtual markets. To what extent fans' consumer behaviour in a virtual environment mirrors that of the physical market, and the impact that extreme behaviour may have on managing virtual service quality are points for further consideration. Our research indicates that fans may in fact display atypical behaviours in relation to the technology and their community of interest.

Music market, supply chain and service quality

Traditionally, there has been a strong correlation between patterns of CD purchase and attendance at concerts, as bigger audiences meant higher record sales for artists

Table 2 A terminology for rock music fans

Hunt et al (1999) Sports Fan	Thorne & Bruner (2006) Popular Culture Fan	Proposed Rock Fan Typology
Temporary Fan Fan for a specific time-bound event such as a home run chase	**Dilitante** Casual involvement with the primary source material. May watch a favoured programme when they can, but will not adjust their lifestyle to do so.	**Casual Fan** • Fan for a limited time, such as coinciding with a chart placing. • May own a small number of recordings. • Makes a casual decision to attend concerts for the object where this fits in with their lifestyle. • May be a regular attender at a broad range of concerts.
Local Fan Bounded by geographical constraints e.g. support local team		
Devoted Fan The object is used to maintain self-concept, so that they are loyal in bad times and attend regularly	**Dedicated Fan** Adjusts lifestyle to pursue interest and actively seeks out other fans	**Loyal Fan** • Band is used to maintain self-concept, with loyalty in bad times such as a weak album. • Identify with the sub-genre of music and attend a focussed range of concerts. • Prioritise attending band's local concert, and may identify themselves by t-shirt wearing. • Have many recordings and some merchandise.

(Table continued)

Table 2 Continued

Hunt et al (1999) Sports Fan	Thorne & Bruner (2006) Popular Culture Fan	Proposed Rock Fan Typology

Fanatical Fan
Move beyond the devoted fan in, for example, going to games in costume. Fan-don is a very important part of self-identification, but at least one other aspect of life (family, work, religion) is more important. Family compliance
	Devoted Fan
	Make major changes to their lifestyle in order to pursue interest and devote a great deal of free time to the activities assocaited with fan-dom. Endevour to become recognised as an expert.
		Die Hard Fan
		• Very important to self-identification over an extended period
		• Will make major changes to lifestyle, but at least one other aspect of life (family, work, religion) is more important.
		• Will prioritise the object to the extent of deriding other bands within the sub-genre.
		• Make considerable investment by travelling to gigs, perhaps going more than once in a tour.
		  • Go to concerts in a t-shirt for a previous tour or dressing up in costume.
		  • Own all or most recordings & collect merchandise
		  • Will be knowledgeable about the back catalogue and may seek opportunities to demonstrate this

Dysfunctional Fan Fandom is primary means of self-identification and is the main method of identification to others. Different from the Fanatical Fan not by degree of self identification, but by anti-social behaviour, ie hooliganism	**Dysfunctional Fan** Engaged in the activity so deeply that they may alienate or become estranged from their family and engage in antisocial activities. Activities may include violence, hysteria or stalking.	**Dysfunctional Fan** • Primary means of self-identification and is the main method of identification to others. • Interference with other aspects of life i.e. 'taking to the road' to follow tours at the expense of work or family life. • May seek contact with the band and occasionally include stalking behaviour. • Will maintain a relationship with others in the community, although may seek to be recognised as an expert by them and may be competitive or antagonistic to the community.

(Connolly and Krueger, 2005). The global recorded music industry is worth upwards of US$38 billion each year, with the market controlled by a small number of dominant record companies. The market for live music consumption is smaller, but statistics from the UK and the US from 2004 to 2006 show that typically around 40% of the population choose to attend a live music event (Waddell, 2006; Ostrower, 2005; Graham *et al.*, 2004; Mintel, 2004). Technology, however, has had a dramatic impact on music consumption, and the relationship where CD purchase was predicated on live consumption is now severed (Plummer, 2006).

The shift in consumption behaviour for recorded music is driven by the increased availability of the internet, and particularly the growing use of broadband, while attendance has been impacted by augmented choices for leisure consumption in the home (Mintel, 2006). Many commentators have predicted that these factors would contribute to a terminal decline, as this comment from Wired magazine illustrates:

> It's not the death of the music industry as we know it, but the funeral isn't far away. (Stroud, 2000).

For the recording industry, the internet has revolutionised the supply-chain by allowing simultaneous richness and reach of information in a difficult market, where consumers are erratic actors, as Graham *et al.* note:

> Record-making is economically as well as technically a complex process [...] Not only is it difficult to identify, develop and manage successful artists, the artistic value (and, therefore, the commercial value) of records depends on their consumers' aesthetic preferences, which are neither stable nor predictable. (2004, p. 1088).

In terms of the live music event industry, in the US, maturation of the market has seen profits focused on a small segment of artists, with the top 1% receiving 56% of concert revenue in 2003, while in 2002, among the top 35 artists, touring income exceeded that of record sales by a ratio of 7.5 to 1. At the same time, the cost to the consumers of purchasing tickets has risen above the rate of inflation for the past 26 years, with an increasing choice of distribution channels in both the primary and secondary markets (Connolly and Krueger, 2005). Clearly, technological innovation has contributed to increased sophistication in the production and consumption of music, and provided new opportunities to capitalise, generate profit and differentiate for competitive advantage. Yet, with these opportunities have come new challenges in terms of maintaining service quality and brand integrity.

The supply-chain for the distribution of tickets for live music events has historically been diffuse, encompassing a range of primary and secondary channels: official venue box offices, phone rooms and internet ticketing agents, as well as brokers, touts and scalpers. Connolly and Krueger (2005) have further identified that the concert industry formerly adopted a casual attitude towards its products however, with the maturation of the market and the dominance of 'big' players – both artists and intermediaries – a heightened professionalism has emerged. As Krueger highlights, 'Early on in the entertainment industry, it's in the interest of the business to think of themselves as throwing a party, not selling a product.' (Plummer, 2006, section 3, paragraph 5).

There are several points of interest here. First, the business model for live music ticketing and the potential impact of internet technologies upon this; secondly, to what extent models of service quality can be applied to the physical and virtual markets for ticket distribution; thirdly, consumer perception and response to ticketing intermediaries and finally, whether changes in the industry generate an attitudinal and behavioural shift in consumers towards their object of fascination – the band itself.

The music industry has a number of discreet yet inter-related areas of operation, of which live performance is increasingly significant for both reputation-building and income-generation. Porter's (2004) discussion of the importance of value-chains in developing and maintaining competitive advantage highlights the importance of scope, inter-relationships and coalitions in a global industry. However, as a dynamic, multi-channel industry, live music events can sit uncomfortably in the linear value-chain model. The main issue is lack of fit with Porter's categories of primary activities, and include the complex nature of the inbound 'logistics', the blurred boundaries between operations and outbound logistics and activities that straddle more than one category such as ticket distribution, which has a strategic resonance beyond the narrow construct of marketing and sales.

The concern is to identify how the Depeche Mode pre-sale offer provides for added value and differentiation, and a value net may provide a better conceptual fit. Value net(work)s can be defined as a set of relationships between firms, where companies engage in multiple two-way relationships to bring increasingly complex products and services to the market (Rajala *et al.*, 2004). Bovet and Martha (2000) further characterise a value-networked organisation as customer-aligned, collaborative and non-linear, digital, fast, and agile.

The inclusion of channel partners in administering the pre-sale points to a networked approach to value: the presence of these actors, both established major players in the recorded and live music industries, is interesting particularly in terms of the perceived and actual quality that they provide. Moreover, the pre-sale represents a consumer-driven or demand-pull offer, as the primary research identifies a clear perception among fans that good seats will sell

out fast. Statistics from Billboard Boxscore (cited by Depeche Mode Message Board postee halo eighteen, 2005), illustrate that 11 out of 21 DM concerts on this leg had 100% sell-out, which appears to buck a trend identified by Connolly and Krueger of declining capacity utilisation for concerts in the US, with only a 75% capacity achieved in smaller venues in 2003 and with arena ticket-sales even lower (2005, p. 19.)

We have previously given consideration to issues of service quality in the music event context (Beaven and Laws, 2004a, Beaven and Laws, 2004b, Laws and Beaven, 2006) and identified that relative levels of intangibility and heterogeneity of music events present the greatest challenges for managers. Consequently, the SERVQUAL model (Zeithaml *et al.*, 1988), which is conventionally utilised to design and measure effective service provision, offers only a partial scale when considering the complex offering of a music event. For this case-study, there is the further dimension of online service quality from the channel partners to consider.

In terms of e-service quality, Yang and Fang's (2004) study of the relationship between online service quality and customer satisfaction identified two inter-related aspects: the design quality of the web-site interface and consumers' perceptions of ease of use and usefulness of this site. They concluded that those factors that generate a negative response would have a greater influence on consumer satisfaction than those having a positive effect. The choice of virtual channel partner could, therefore, be the key in maintaining customer satisfaction.

For the DM pre-sale, fans were required to access the offer via iTunes and Ticketmaster (TM). While perception of the iTunes brand and its service appear from our research to be fairly benign, Ticketmaster received some strongly worded criticism. Anecdotally, this attitude is not limited to the particular fan-set as TM's dominance of the entertainment ticketing market in the US and the UK, and its close commercial relationship with major concert promoters such as Clear Channel Entertainment (now known as Live Nation), have been criticised for creating a monopoly (Connolly and Krueger, 2005). In the UK, TM has recently been considered in an Office of Fair Trading investigation into competition for ticketing distribution, which found no evidence of unfair business practice (OFT, 2005). It is apparent that most consumers are unaware that TM receives no income from the face value of the ticket, and that a proportion of its service charge may also be passed to the venue or promoter. Indeed, TM have been found in several instances to charge lower fees than their competitors for identical tickets (OFT, 2005, pp. 29–30). Nevertheless, the perception of TM as a 'necessary evil' appears to persist in the blogosphere and print media, as the company's dominance effectively restricts consumer choice, despite claims by TM that the service adds value (Emmons, 2000).

Evidently, profit margins for ticket distribution agents in the live music events industry are narrow, and, therefore, the focus for competitive advantage resides in the volume of market share. In this way, TM's prioritising of industry-facing business-to-business (B2B) activities rather than business-to-customer (B2C) service becomes explicable, although our research would suggest that this is a short-sighted attitude for the development of value nets in the virtual marketspace.

So, the interdisciplinary literature that informs this study suggests a substantial shift in the concert and music businesses coupled with increasing intensity of fan attachments that produces not only opportunities but challenges in the live events sector. This study, therefore, focuses on ticketing as a critical interface between business and customer that is illustrative of wider concerns.

METHODOLOGY

As a means of exploring this new and developing area of live event ticketing, we have drawn on data from the Depeche Mode Message Board (DMMB) as a rich source for accessing relevant market segments. We analysed the data set through a research design drawing on netnography, an approach based on interpretation of computer-mediated textual discourse, which differs from ethnography by the obscured identities of participants and the public nature of discussions (Kozinets, 2002).

The netnography methodology was developed by Robert Kozinets in the 1990s (1999, 2002) for conducting online marketing research. With reference to the relatively small number of subsequent studies to apply netnography (Nelson and Otnes, 2002, Maclaren and Catterall, 2002, Yang and Fang, 2004, Langer and Beckman, 2005, Morgan, 2006), we made some modifications to Kozinets' approach, most notably with respect to the issue of consent of the IMB members whose posts we reviewed, and did not participate in the discussion through either public posts or private email messaging. Although we discuss the detail of this process elsewhere (Beaven and Laws, 2006), it is pertinent to outline our reasoning here. Kozinets (2002) noted that no clear consensus on the public or private nature of message boards had yet emerged and that the requirement for consent was not clear. He stated, therefore, that ethical considerations determined that consent should be sought. Langer and Beckman (2005) develop an argument for conducting netnographic studies without informing the subjects or seeking consent for sensitive topics, as their own research shows that the intervention of the researchers can interfere with the findings. We do not claim our topic is a sensitive one, but drawing on the successful outcome of the Langer and Beckman study, and the increasing 'normalisation' of e-communication since Kozinets developed the methodology, we assert that IMBs are now established as public forums of communication and that consent has become unnecessary for analysis of public postings. Our research design therefore took the following form: identification of the data set, a period of immersion in the data to familiarise ourselves with the findings, qualitative textual analysis to develop a narrative, quantitative analysis of the posts to determine postee typologies and orientation to the community and an extreme case analysis to determine the boundaries of the discussion.

In addition, of particular interest to us was the work of Yang and Fang (2004), Morgan (2006) and Stubbs (1999) in establishing a 'lifecycle' for an issue-based IMB debate as around 10 or 11 days, reflecting the accelerated speed of communication, which the medium not only facilitates but seems to demand, and coinciding with the results of our study.

IMBs are sites for asynchronous chat on the internet where, often using a pseudonym, people can post relatively anonymous messages and replies grouped into themes or threads. Registered users have the option to present information about themselves, such as gender, age, occupation and geographical location, to IMB visitors; this information may or may not be accurate. In addition, the IMB automatically generates, and makes available to any viewer, information about the rate and volume of posting by each user. This study made reference to these 'user profiles' and to the posting history, in addition to the analysis of the content of posts, in the characterisation of the postees that are referred to in the findings section of this paper.

We suggest, through drawing on the information about each user's posting history to the DMMB that is made publicly available on the message board, that this data set is indicative of fans we have already characterised as loyal, die hard or dysfunctional, and provides a fascinating and detailed window on their experiences and responses.

IMBs provide valuable feedback for artists and marketers, and promote a community among users. Although 'online communities present fairly explosive environments', (Kozinets, 2002, p. 65) the chat is generally regulated, with users required to register and with posts subject to moderation and censorship. The DMMB is the 'official' message board for fans, and is linked to the band's own website. Although by no means is it the only IMB for Depeche Mode fans, comparison with other key sites as set out in Table 3 demonstrates that it is by far the largest. At the time we collected the data, on 21 September 2005, the DMMB had 22,501 registered members who had contributed to 12,712 threads with 271,138 posts, since the establishment of the IMB several years earlier. It can, therefore, be characterised as an active IMB with a complex community of registered users.

The case study covers the Ticketmaster pre-sale for the first US-leg of the Depeche Mode 2005–2006 world tour. This was run by allocating a unique pre-sale password to anyone within the USA who ordered a digital download of the album *Playing the Angel* from iTunes during the offer period. The password was used to access pre-sale tickets.

The chronology of the presale was as set out in Table 4.

For the purpose of this study, we reviewed 24 threads related to the pre-sale with 485 posts by 144 postees. Posts were made over a ten-day period from the announcement of the discovery of the pre-sale at 11.23 pm Mountain Standard Time (MST) on 12 September until 21 September 2005, after tickets had gone on sale and participating members had purchased their tickets. The rate of posting compares favourably with Langer and Beckman's study (2005), where 896 posts

Table 3 Depeche Mode IMB Comparisons

	DMMB	Empty World 3	Depeche Mode Forum	Depeche Mode.TV	Indigo	DepMod
Members	100%	62.2%	13.2%	9.2%	2.6%	2.3%
Threads	100%	71.5%	4.3%	9.2%	n/a	14.4%
Posts	100%	49.9%	3.8%	8.9%	0.9%	8.6%

Sources: quoted 3rd July 2006: www.depechemode.com, www.depeche-mode.com, www.depechemode-forum.com, www.depechemode.tv, www.depeche-mode-world.de, www.depmod.com

were gathered over a 16-month period. However, the relatively short timeframe meant that we were unable to apply Kozinets' categories for the level of involvement of individuals in the community (2002), but analysis of the postees led to the identification of 28 key community members, selected because they had either initiated threads, contributed detailed and expert-like posts or were high-volume contributors during the period under review. Most of those falling into the final category were also found to be high-volume contributors to the DMMB over an extended period, for example the automatically generated post data for Ilse101 shows more than 3,000 messages over a three year period, with an average count of 3.25 posts per day. However, the post-count of some 'experts' suggest a weaker connection to the IMB community, with Hal9000 making an average of 0.07 posts per day, since joining the DMMB in 2005.

Posts ranged from just one line to more than 600 words, although most posts were substantially shorter. We have in each case replicated the original language, syntax and grammar of the postee, despite the occasional inconsistency that this creates within our text. Many of the posts contain expletives, which we feel demonstrate the strength of opinion and are integral to the argument, and we, therefore, took the decision to retain them. All times and dates quoted are Mountain Standard Time (MST) or Greenwich Mean Time (GMT) −7, and we have adopted the US date format.

FINDINGS

In this section, we present the case study as a chronological narrative, drawing out themes that developed from the discourse, and relating these to our fan typologies. We will go on to analyse and generalise

Table 4 Pre-Sale Chronology

Date	Pre-sale stage
16 June 2005	Spring 2006 European-leg dates released and free presales begin.
8 September 2005	Autumn 2005 US-leg dates released.
12 September 2005	US presale on iTunes shortly before the official announcement was due. Begins to 'leak' to the message boards.
13 September 2005	iTunes begins taking pre-orders of Playing the Angel. Anyone ordering is promised their password within 48 hours.
20 September 2005	Original date for commencement of ticket presales.
21 September 2005	Revised date for commencement of ticket presales.
24 September 2005	General release of tickets to the public.
28 October 2005	First date of US leg of Touring the Angel.

themes for event managers in the concluding section.

Perceptions of the Value of this Novel Approach to Pre-sale

Allocation of tickets

We found that there was a great deal of ambiguity amongst DMMB postees about the appropriateness of the pre-sale format, some of which is related to fan typologies. Although, as predicted by Kessler (BBC News, 2005), there was enthusiasm by some about the use of cutting edge technology, there was confusion and even distress from others about their inability to cope with the technological aspects. For one group, there was a much more fundamental concern that the concept was exploiting their high degree of loyalty to the band:

> While digital media is cool and all, this **bites**, because it means all of us who prefer to purchase the pressed CD – for longevity and a full printed insert – get the short end of the proverbial stick … Oh, I can smell the dineros that changed hands for this deal…☹
> [Ocelot, posted 2.45 am
> MST, 13 September 2005]

This was the dominant theme of the discourse in the phase prior to the pre-sale. Posts included pejoratives such as 'shakedown', 'gimmick', 'sell out' and 'insult' to describe the offer, and repeatedly questioned why their loyalty was being ignored. A key DMMB member N2theBlu said in initiating the thread 'Grabbing Hands Indeed! iTunes Coercion':

> The fans here who have shown their devotion to the band by joining this site (following the band's progress, participating in forums, etc…) should be the ones rewarded with pre-sale privileges, not the masses who blindly gulp down whatever the music industry marketing machine decides to force-feed them.
> [N2theBlu, posted 10.16 am
> MST, 13 September 2005]

The overwhelming majority of postees suggested they would be purchasing the standard CD format in addition to the download offer, because they were collectors and wished to own the full sleeve art. We have identified this as a feature of the die-hard fan, and for most, this was characterised not only as a 'badge' of their ultra-loyalty, but they suggested it was a predictable action that could be relied upon by the band and their management. Many questioned the practice of 'pressurising' them to buy the album twice as an attempt to manipulate the chart placing, including Songwritingguy who, despite being a casual member of the DMMB community, presents himself as a music industry insider and a serious fan:

> Being the die-hard fan that I am, I'd have to do the pre-sale thing to get a "chance" for tickets and then buy the commercially available copy to have as part of my collection. That's a double-dip and a multiple of two for the Soundscan [chart] sales figures.
> [Songwritingguy, posted 9.09 pm
> MST, 13 September 2005]

These serious fans were split between those intending to participate unwillingly, referring to feeling let down, and those displaying dysfunctional characteristics (where the process of fandom overtakes the fascination with the object), who said that they were so angry or disappointed that they would not take part in the pre-sale offer:

> I never complain about Mode …but this has me livid! I love having the cover art and the actual CD in my hand. And, NO, I don't want to buy both! And that's the ONLY way I can participate in the presale???!!?? Screw it, I'd rather give a broker my $, at least they're honest that that's all they want from me.
> [MartinsFluffyHead, posted 1.20 pm
> MST, 13 September 2005]

Others who might be characterised as serious fans, however, were not concerned and felt the offer was a fair or an attractive one. Repeated comparison was made by this group with the presales offered by other bands linked to memberships of their fan clubs. Postees commented that for bands such as the Rolling Stones, access to the presale was through a fan club at a cost of up to $100. For this group, the concept of paying to access a presale was not a controversial one and the focus was on the cost of the offer, as this typical post suggests:

> I too prefer the actual CD. But I think it's well worth the $12 if it will get me pre-sale tickets and good seats at the show. So yeah, I'm paying double for the iTunes and the CD, but so what?
>
> [Devoted1 AA, posted 9.47 am MST, 13 September 2005]

For a small number of postees, however, we found that the cost of the deal along with ticket commission, was an issue that was likely to determine whether or not they participated. This post from Melissa1380 is interesting because she has highlighted an issue which runs through very many of the discussions we reviewed: she is resistant to the shift in ticket distribution channels for rock concerts that this deal exemplifies:

> I've done the TicketMaster thing and [with] all the charges, it ends up pretty pricey. Do you think pre-sales is really worth it, or would you just take your chances and buy them (I'm from Detroit and can buy them right around the corner) as soon as they go on sale.
>
> [Melissa1380, posted 1.36 pm, MST, 12 September 2005]

We found that many of the mature and established concert-goers resented the move away from having a choice to queue up at venues and buy tickets face-to-face from the box office. The increased pressure of online pre-sales and the wider ease of ticket purchase via websites and agents such as TM has taken away their choice to purchase from a venue. We detected a palpable nostalgia for this to the extent that it is perceived

as an integral part of the 'true fan' experience. For others, this was anachronistic:

> I find it somewhat surprising that the same people who would wait 36 hours in line to buy tickets, or be willing to pay a hell of a lot more to meet the band, complain about $12? Even if you only earn minimum wage missing a day of work is about $40. winge winge winge.
>
> [Dismalgrrrl, posted 6.26 pm MST, 13 September 2005]

As the pre-sale process commenced, views about ethics and the value of the deal shifted rapidly from day to day, and we return to this issue further on in the paper.

While US fans able to partake of the offer debated its merits and whether to participate, it was evident that some die hard or dysfunctional fans from outside the US were desperate to be allowed to take part, but were unable to purchase from the US iTunes site. Not expecting a pre-sale 'closed' to the USA, some had already booked airline flights and hotels before the details were announced, and were frustrated at not being able to have a first chance of getting the best tickets. This discussion is indicative of the high levels of loyalty evident among this group, and of the global nature of fandom.

Another sharply felt issue was the activity of ticket touts or scalpers, who re-sell concert tickets for sums potentially considerably above face value. Results of the study and evidence from the literature suggest that the activity of touts has accelerated as the internet has become the premium distribution channel for tickets for major rock gigs, and this is an issue meriting further research. We found a great deal of innuendo and suspicion among the postees about collusion between official agents and touts, and many serious fans certainly felt that going to a tout was the only way of buying good seats for arena gigs. The chance to thwart the tout or scalper was initially seen as a major advantage of the pre-sale, this quickly changed when postees realised that the download fee was a tiny investment and would not impede the participation of touts in the pre-sales. This was confirmed by the appearance of tickets on e-Bay on the first day of sale.

The buying process

So far, we have concentrated on issues relating to perceptions of the value of this novel approach to pre-sale allocation of tickets, focussing on data collected in the early part of the study before the offer was implemented. We will now concentrate on the operations and delivery of the offer, and the attitudes of serious fans to service delivery.

As suggested earlier in the paper, despite all subjects in the study being users of web-based technology, there was some confusion or anxiety about the technological aspects of the offer. The first stage of the process for participants was to pre-order a copy of the album Playing the Angel from iTunes, and some postees expressed confusion about the difference between iTunes and an iPod, for example:

> Does anyone know if you have to actually have an iPod to download the album? I am willing to pay the $12 if it means I can participate in the pre-sale, but I don't have an iPod.
> [Elektra23, posted 12.43 pm MST, 13 September 2005]

This dovetailed with concerns about the price of the pre-sale, as some people wrongly assumed that they would need to additionally purchase an iPod. Furthermore, there appeared to be a trend to reject technological developments, particularly the purchase of music by download, because this challenged their notion of authenticity and norms of behaviour for 'true' rock fans. This group tended to amplify the difficulties related to the distribution channels and made regular references to the established

past practice, such as queuing up for tickets or buying vinyl.

On 13 September 2005, one day after the pre-sale announcement, fans were able to begin pre-ordering the album from iTunes. Having pre-ordered – a process that caused some deal of confusion – participants then had to wait for their unique password to be emailed to them by TM. This was promised within 48 hours. Many postees demonstrated a typical assumption that e-transactions will be instantaneous or near-instantaneous by becoming very concerned when their passwords took hours or even days to be delivered. This proved to be a critical moment in terms of customer perception and satisfaction, and we noted an increase in posting activity during this period. We also noted that some key DMMB members began to focus frustration away from TM and towards the band and their management:

> It sucks but considering the piss poor organization that [the band's manager Jonathan] Kessler used when booking the tour... The USA tour is simply the warm up gig...they'll work out the bugs here and Europe will get the killer shows and most likely better set lists.
> [Crewboy, posted 6.56 pm MST, 13 September 2005]

While the terms of the offer were that TM were due to deliver passwords within 48 hours, some postees become highly agitated and began regularly checking emails well before this time had expired, including TJ tank who is a high-volume postee and thread initiator:

> I'm kind of nervous! I pre-ordered yesterday around this time (9:00am) and still haven't gotten anything. I've checked my email about a dozen times so far today.
> [TJ tank, posted 8.07 am, 15 September 2005]

Frustration soon developed into telephone calls and complaints to TM. Posts from the webmaster and other postees explaining the 48 hour turnaround did little to calm nerves. We note that this level of agitation was not necessarily confined to the dysfunctional typology, of whom we note very few in the study, but was an unexpectedly acute reaction from fans who might be characterised as both die hard and loyal. People calling TM but receiving a poor customer service response were responsible for posting a string of increasingly angry and abusive comments during an intense period of board activity on the 13, 14, 15 and 16 September 2005.

> Yea, I jumped on the presale this morning when I saw it, and I haven't recieve anything from TicketMaster about a presale password. I'm starting to think I've been had.
>
> [Mande, posted 11.09 am MST, 13 September 2005]

> I have also been playing this silly ticketbastard game...and i Have been running the gauntlet of customer service with tm and itunes
>
> [Darkspark, posted 9.28 am MST, 16 September 2005]

> ☹ WTF? They are taking the p!ss! See, I've already been thru this w/ them before, on a much SMALLER SCALE, for the Exciter tour. And even then, I was thinking, "THIS IS TOO MUCH TROUBLE TO BE GOING THROUGH JUST TO GET BAD SEATS TO A CONCERT!!!" ☹ And then I said, "Never again! ☹ I won't go thru this again." So I won't. I just won't bother.
>
> [LilyKiss, posted 9.47 am MST, 16 September 2005]

Concerns included representatives of TM who were unaware of the pre-sale or the tour, had no knowledge of the password allocation systems or gave standardised responses to customers. It is notable, therefore, that during this burst of anxiety about poor service delivery, many of the issues that came up repeatedly in posts were related to TM's 'traditional' telephone operations, and not to the online service. Among the many posts during this four-day

period, one of the less abusive that we found was this:

> Well, calling TicketMaster will get you no where... they didn't know anything about ANY of the DM concerts that are going on in the states- presale or regular sale. I read the complete email I got from TicketMaster to two different people, one a regular phone person, the other a manager. The manager was really confused why I had gotten such an email since she knew nothing of the promotion...argggggggggggggggggggg. Get your acts together already.
>
> [Housegirl, posed 7.14 pm MST, 14 September 2005]

Whereas another, who we would categorise as a casual fan, took the opportunity to post repeatedly on the same aggressive theme:

> As fucking usual, the assholes at TicketBastard don't know their heads from their asses. I pre-ordered the new album on iTunes. Have I gotten an email with the pre-sale password? No. I called TicketBastard, and after the usual runaround, they say they don't know jack-shit about any pre-sale or passwords, even after I pointed them directly to their own web-page... Freaking morons. Someone needs to do their fucking job and fix this. I'm so freaking mad that I wish a hurricane or terrorists had done away with TicketBastard HQ.
>
> [Hal9000, posted 2.00 pm MST, 13 September 2005]

Just a few voices were heard attempting to moderate this aggression towards the call centre staff, including this post from an initiator:

> Please don't vent your frustration on the person answering the call, most likely they know nothing about this as it was NEVER communicated to them. I am not a big fan of TicketMaster but I work on a help desk and anytime changes are made we are the last people to know.
>
> [Windscreen, posted 8.28 am MST, 15 September 2005]

Finally, on 16 September 2005, after a great deal of continuing rumour and speculation, the first DMMB members began posting around midday to announce that their passwords had arrived. Whereas this apparently did little to improve the mood of the recipients of the passwords, inevitably it simultaneously heightened anxiety among those still waiting. Furthermore, a significant error became apparent at this time: there was a 24-hour discrepancy between the stated start time for the pre-sale on the official website (20 September) as opposed to the email from TM (21 September). For a fast moving ticket distribution channel this time difference could be critical and the commencement of the pre-sale had to be delayed.

Reflections on the value of the offer

The final phase of the offer was the sale of tickets, which began belatedly on 21 September 2005. Mirroring the extreme excitement and anxiety we found throughout the lead-up phase, DMMB members were online before the pre-sale opened, swapping stories and tips about how to get the best tickets, sharing and accentuating their experience.

As soon as the presale began for east coast shows, members began posting details of their ticket purchases. Many of the messages were very positive in nature, with fans swapping congratulations for securing good seats or expressing their excitement:

> 7th row for Tampa!!!! Never in my life have I ever had tickets this good - and I didn't think I ever would. My level of excitement is just beyond words.
>
> [Firmbeliever, posted 7.01 am MST, 21 September 2005]

> Was it all a dream? Did we really score great floor seats w/o having to pay scalpers hundreds of dollars? Somebody pinch me!
>
> [Bean, posted 6.30 am MST, 21 September 2005]

> I'm a buzzing with excitement right now. I will finally be able to see Depeche Mode without binoculars!
>
> [Sunlovey, posted 6.20 am MST, 21 September 2005]

Other superlatives include 'Great day in Depeche Mode Land!!!', 'Thank you Depeche Mode, Itunes, TicketMaster!!!! I love you all!!!!' and 'I'm SOOOOOOO excited, I can't quit shaking!!!!!!' Not all postees, however, were happy with the process. There was an extensive, urgent and fast-moving discussion in several threads from fans searching for better seats and discussing tactics for securing them. Others were very unhappy with their allocation:

> How did you guys get such good seats. I was on at 10:00 sharp and I got section 118 row d!! needless to say I am upset. Back in the day I waited in line for DM tix and always got great seats!
>
> [Yves naumann, posted 6.18 am MST, 21 September 2005]

Finally, during the morning sales began, postees were very quick to reflect upon the offer and react to the allocation of tickets. Here, it became clear that despite previous posts complaining of being manipulated by the offer and of poor customer service delivery, the primary concern of the majority of postees was the quality of the tickets they were able to purchase. In this respect, responses were almost universally positive – with some extreme excitement demonstrated by fans at the dysfunctional end of the typology spectrum. This mood is summed up by this 'thread-opener' from a high-volume poster who had previously been highly critical of many aspects of the offer:

> Let it be known, the pre-sale ROCKED. It came through in a major way and was WELL worth the $12 spent. If you haven't taken advantage of it, do. My sincere thanks to everyone who put this thing together. It was nerve racking getting to this point, as

they ironed out the kinks but they came through.

[Sunlovey, posted 7.16 am MST, 21 September 2005]

CONCLUSIONS

The data presented here maps the experiences and responses of loyal fans of the rock band Depeche Mode to a radical shift in the pre-sale offer for the US leg of their 2005–2006 World Tour. The offer illustrates innovation through its utilisation of technology, and is promoted as a value-added service to fans by the band's management. As such it may provide a model of practice for the wider festival and event sector in the future, and indeed the approach has already been adopted by other established rock acts such as the Red Hot Chili Peppers. This research study has identified two clear areas for consideration in judging the efficacy of new ticket distribution channels for live music events: the organisation of industry channel partners for efficient supply, and the management of the relationship with the fan community who are principal purchasers of the product offer.

The music industry has been identified as pioneering B2B and B2C interactions in virtual marketspaces (Weiber and Kollmann, 1998; Woerndl *et al.*, 2005) but there has been a steep learning curve towards achieving high standards of virtual service quality. The requirement of multiple channel partners to facilitate a pre-sale presents additional layers in the otherwise direct relationship between an individual fan and their object of fascination and, if not actively monitored, may create an opportunity for brand pollution of the band itself. As consumers of live events become more familiar with technological developments, organisations must act more strategically to address their needs, and to negate the potential impact on customer satisfaction of poorly conceived, overly-complex and under-resourced e-services.

Ticketing for major rock acts, festivals, high profile and special events has, *de facto*, shifted to on-line transactions as the speed of purchase has increased. Promoters may offer other distribution channels, but these are redundant if tickets sell out quickly through online agents. Evidence from the literature (Delgado-Ballesster and Munuera-Alemán, 2001) and our findings suggest that, while the market may stand an increase in the face value of tickets well above inflation, there is significant resentment about extras such as agent's commission. This resentment is linked to the tolerance of consumers' brand trust for change and also ties in with concerns about service quality for online transactions. We sense an assumption by the channel partners in this study that virtual transactions warrant less care. However, we have demonstrated that, far from being a functional transaction, the ticket buying process is part of the ritual of rock concert and festival attendance, and, therefore, integral to the relationship between fan and object. Research by Hunt *et al.* (1999) indicated that the degree of fan loyalty is predicated on perceived brand performance fit, and channel partners need to demonstrate greater awareness in planning virtual marketspace strategies to limit alienation in the purchase process.

Furthermore, we have highlighted an ambiguity towards technology rooted in the norms of the genre and attitudes to authenticity, which predisposes some rock fans to negative attitudes towards online ticket purchase. TM has demonstrated technological innovation in the primary and (since 2005) secondary distribution markets, which is intended to bring positive benefits to the consumer. However, frustration with the low quality of service in TM's call centres post-purchase can outweigh these benefits, which fits the Yang and Fang (2004) model for effective customer satisfaction.

e believe that the scale of TM's market dominance is at the heart of dissatisfaction, as the corporate business model and the ease of purchase intercede with fans' direct relationship to the band. The traditional development of the fan relationship through investment of time and effort has been annulled through technology and from the evidence of this study, appears to increases 'luddite' behaviours and criticism from serious fans that others are not paying their dues. This 'stubbornness' is combined with a nostalgic yearning for a time when music was a less aggressively commercial enterprise. This would tie in with attitudinal shifts such as the need for authentiseeking identified in the literature. It also indicates the self-effacing nature of fandom, where the fan typically absents themselves from any consideration of their relationship with the object (Hills, 2002): perhaps the overtly commercial nature of the transaction is a sharp reminder of the fans' relative position in the value-exchange structure of fan consumption. The question of whether this response is age-related may warrant further study, however we suggest that as our research population were all IMB users, the manifestations of irrationality and authentiseeking in the findings should not be confused with either a lack of technical prowess or unwillingness to use the Internet, *per se*. However, analysis of demographic factors falls outside the scope of this netnographic analysis although Kozinets (2002) has identified that netnographic study can be further developed with triangulation, and this is a point for further consideration.

Another key area for ensuring consumer loyalty is communication. To achieve high levels of satisfaction, consumer expectation must be managed within the value chain and originators have to take some responsibility for ensuring quality and flow of information down stream. A danger with virtual communities such as IMBs, away from the 'official' information channels, is the tendency for rumours and panics to spread quickly, as people whip themselves into frenzy, accentuated by the blurred distinction between official and unofficial information. This contrasts with the traditional PR model of an official press-release followed by a controlled number of contact-points, and has heightened the need for greater clarity and clear information from the start of an offer. Furthermore, the 'click and get' mentality of virtual markets creates new challenges around speed of response, and we have illustrated ways in which combining virtual and real products is problematic, when attempting to maintain a balance between supply and demand.

Effective partnerships for the future of festival and event management in the marketplace/marketspace are situated in the adoption of a value-net approach, which has been described as consumer-driven and flexible. However, this study indicates that many of the drivers in a pre-sale are industry-focussed. In a market where loyal fan business will follow the artists, the business to be 'won' is B2B. For a fan, if their favourite band are promoted by a faceless major player, perform in what they perceive to be characterless barns, and sell tickets through an agent they feel is ripping them off, they may be convinced they are dancing with the corporate devil, but do they REALLY decide they won't buy the tickets to experience the show?

Analysis of posts to the DMMB have illustrated that many fans regard the band as above reproach and apportion blame for the poor fit of the pre-sale offer to the other channel partners. However, this perception of the band as an unsullied artist is (deliberately?) naïve, as bands of long-standing have evolved well beyond this career stage: for Depeche Mode, we identify that this transition to 'band as brand' occurred on the back of the Music for the Masses album, which broke them in the US after 1987

(Pennebaker, 1988). Depeche Mode have, therefore, successfully negotiated the journey from 'party-throwers' to professional musicians in terms of how they manage their business; yet, the service journey that their fans must continue to negotiate is less well resolved, and developments in marketing Depeche Mode's product for consumption have actually obfuscated the relationship from the fan perspective. Clearly, the virtual space for music production and consumption that internet technologies have opened up is a contested one, where identities and relationships are currently being re-negotiated.

Traditional forms of interaction and established patterns of behaviour between industry and consumer are now being challenged (Murray, 2004; Jones, 2000). Stratagems such as the Depeche Mode pre-sale represent an attempt by the industry to respond to the new marketspace for music consumption. Despite claims by those closely connected with the band that this is about rewarding fan loyalty, the lack of attention to all stages in the distribution and delivery of the pre-sale indicates that the full implications of the shift for managing the fan–object relationship in a positive manner have not yet been grasped.

At this point, it can be stated that fans' consumer behaviour with regard to channel partners should remain a priority for the artist and promoter as the extremity and non-rationality of their reactions could undermine loyalty in the long-term. In the live music event market, where concert attendance is now the main source of revenue for artists, lack of attention to this kind of detail is a serious oversight.

In the introduction we outlined that research into event management must involve a practical application in order to have currency. As the music industry continues to move swiftly towards virtual marketspace models for interaction, structured reflection on the practicalities of managing

that process need to be undertaken. Furthermore, factors noted here, particularly the shift in consumer expectations as a result of changes in communication channels and the as-yet limited response of the industry, has a resonance beyond the music industry. For managers of a range of festivals, concerts and other events, the notion of what constitutes good events management must now encompass the totality of the event experience across the virtual and physical domains.

ENDNOTES

1. © M.L. Gore (1987) published by Grabbing Hands/Sonet.

REFERENCES

Arts Council of England (ACE)/Henley Centre (2000) Towards 2010: new times, new challenges for the arts, London, Arts Council of England.

BBC News (2005) Depeche Mode ticket download plan, 14 September, 2005, available: http://news.bbc.co.uk/1/hi/entertainment/music/4244974.stm, accessed 14 September 2005.

Beaven, Z. and Laws, C. (2004a) Just the Ticket? *Arts Professional*, **79**, 5.

Beaven, Z. and Laws, C. (2004b) Principles and applications in ticketing and reservations management. In I. Yeoman *et al.* (eds) *Festival and Events Management: An International Arts and Culture Perspective*. Oxford, Elsevier Butterworth-Heinemann, pp. 183–201.

Beaven, Z. and Laws, C. (2006) 'World in My Eyes': An evaluation of netnography as a methodology for audience studies. Paper presented at the LSA 2006 Conference 'Making Space: Leisure, Tourism and Renewal', 11–13 July 2006, University of the West of England, Bristol.

Bovet, D. and Martha, J. (2000) Value nets: reinventing the rusty supply chain for competitive advantage, *Strategy & Leadership*, **28**(4), 21–26.

Connolly, M. and Krueger, A. B. (2005) Rockonomics: The economics of popular music,

NBER Working Paper No. W11282, available at: http://econ-www.mit.edu/events/pdf.php?id=1123, accessed 29 June 2006.

Delgado-Ballesster, E. and Munuera-Alemán, J.L. (2001) Brand trust in the context of consumer loyalty, *European Journal of Marketing*, **35**(11/12), 1238–1258.

Deller, J. (2006). A La Mode. *Observer Music Monthly*. 15 October 2006.

Depeche Mode Message Board (2005), available at: http://forums.depechemode.com/forum/, accessed on various dates from September, 2005–June, 2006.

Emmons, N. (2000) Electronic Ticketing Continues Growth Pattern, *Amusement Business*, **112**(37), 17, available at: EBSCO Business Source Premier, accessed 22 June 2006.

Graham, G., Burnes, B., Lewis, G. J., and Langer, J. (2004) The transformation of the music industry supply chain: a major label perspective, *International Journal of Operations & Production Management*, **24**(11), 1087–1103.

Gyimóthy, S. (2000) Odysseys: analysing service journeys from the customer's perspective, *Managing Service Quality*, **10**(6), 389–396.

Hayes, D. and Slater, A. (2002) 'Rethinking the missionary position – the quest for sustainable audience development strategies, *Managing Leisure*, **7**(1), 1–17

Hills, M. (2002) *Fan Cultures*, London, Routledge.

Hunt, K. A., Bristol, T. and Bashaw, R. E. (1999) A conceptual approach to classifying sports fans, *Journal of Services Marketing*, **13**(6), 439–452.

Kozinets, R. V. (1999) E-Tribalized marketing?: The strategic implications of virtual communities of consumption, *European Management Journal*, **17**(3), 252–264.

Kozinets, R. V. (2001) Utopian enterprise: Articulating the meanings of Star Trek's culture of consumption, *Journal of Consumer Research*, **28**(1), 67–88.

Kozinets, R. V. (2002) The field behind the screen: using netnography for marketing research in online communities, *Journal of Marketing Research*, **39**, 61–72.

Langer, R. and Beckman, S. C. (2005) Sensitive research topics: netnography revisited, *Qualitative Market Research: An International Journal*, **8**(2), 189–203.

Laws, C. and Beaven, Z. (2006) Practice/policy dichotomies in the successful delivery of art events: an operations management perspective, in: S. Fleming and F. Jordan (eds) *Events and Festivals: Education, Impacts and Experiences*, LSA Publication No. 93, Eastbourne, Leisure Studies Association.

Leppert, R. and McClary, S. (eds) (1987) *Music and Society: the politics of composition, performance and reception*, Cambridge, Cambridge University Press

Maclaren, P. and Catterall, M. (2002) Researching the social Web: marketing information from virtual communities, *Marketing Intelligence and Planning*, **20**(6), 319–326.

Miller, J. (2003) *Stripped: Depeche Mode*, London, Omnibus Press.

Mintel. (2004) Music concerts and festivals – UK. Mintel Group Ltd, available at: http://reports.mintel.com [Accessed: 29 June 2006].

Mintel. (2006) Leisure Time – UK – February 2006. London: Mintel International Group Ltd, available at: http://reports.mintel.com, accessed 30 June 2006.

Morgan, M. (2006) Festival spaces and the visitor experience, Paper presented at the LSA 2006 Conference 'Making space: leisure, tourism and renewal', 11–13 July, 2006, University of the West of England, Bristol.

Murray, S. (2004) 'Celebrating the Story the Way It Is': Cultural Studies, corporate media and the contested utility of fandom, *Continuum: Journal of Media and Cultural Studies*, **18**(1), 7–25.

Nelson, M. R. and Otnes, C. C. (2002) Exploring cross-cultural ambivalence: a netnography of intercultural wedding message boards. *Journal of Business Research*, **58**, 89–95.

Ostrower, F. (2005) *The Diversity of Cultural Participation: Findings From a National Survey*, Washington DC, The Urban Institute.

Pennebaker, D. A. (1988) *Depeche Mode 101: A film by D. A. Pennebaker*, [VHS Video], London, Mute Film.

Pine, J. and Gilmore, J. H. (1999) *The Experience Economy: Work is Theatre and Every Business a Stage*, Boston, MA, Harvard University Press.

Plummer, R. (2006) Winners take all in rockonomics. BBC News, 20 April 2006, available at: http://news.bbc.co.uk/go/pr/fr/-2/hi/business/4896262.htm, accessed 29 June 2006.

Porter, M. E. (2004) *Competitive Advantage: Creating and Sustaining Superior Performance*, New York, NY, Free Press.

Rajala, R., Westerlund, M. and Rajala, A. (2004) *Business Models and Value Nets as the Context of Knowledge-Intensive Service Activities in the Software Business*, Helsinki, LTT Research Ltd.

Redden, J. and Steiner, C. J. (2000) Fanatical consumers: towards a research framework, *Journal of Consumer Research*, **17**(4), 322–337.

Roberts, K. (2004) *The Leisure Industries*, Basingstoke, Palgrave Macmillan.

Ross, K. and Nightingale, V. (2003) *Media and Audiences: New Perspectives*, Maidenhead: Open University Press.

Sandvoss, C. (2005) *Fans: The Mirror of Consumption*, Cambridge, Polity Press.

Silvers, J. R., Bowdin, G., O'Toole, W. J. and Nelson, K. B. (2006) Towards an international event management body of knowledge (EMBOK). *Event Management*, **9** (4), 185–198.

Stroud, M. (2000) A Music Industry Death Knell? *Wired*, 11 January, available at: http://www.wired.com/news/culture/0,33559-0.html, accessed 29 June 2006.

Stubbs, P (1999) Virtual Diaspora?: Imagaining Croatia On-line, *Sociological Research Online*, **4** (2), available at: www.socresonline.org.uk/socresonline/4/2/stubbs.html, accessed 29 June 2006.

Thorne, S. and Bruner, G. C. (2006) An exploratory investigation of the characteristics of consumer fanaticism, *Qualitative Market Research: An International Journal*, **9**(1), 51–72.

Waddell, R. Touring pulls out of slump. Billboard Magazine, April 29, 2006, available at: www.billboard.biz/bb/biz/archivesearch/article_display.jsp?vnu_content_id=1002345285, accessed 29 June 2006.

Weiber, R. and Kollmann, T. (1998) Competitive advantages in virtual markets – perspectives of "information-based marketing" in cyberspace, *European Journal of Marketing*, **32**(7/8), 603–615.

Woerndl, M., Powell, P. and Vidgen, R. (2005) Netsourcing in SMEs: E-ticketing in Art Venues, *Electronic Markets*, **15**(2), 119–127.

Yang, Z. and Fang, X. (2004) Online service quality dimensions and their relationships with satisfaction: A content analysis of customer reviews of securities brokerage services, *International Journal of Service Industry Management*, **15**(3), 302–326.

Zeithaml, V. A., Parasuraman, A. and Berry, L. L. (1988) *Delivering Quality Service: Balancing Customer Perceptions and Expectations*, New York, NY, Free Press.

Glasgow's Winter Festival: Can cultural leadership serve the common good?

Malcolm Foley and Gayle McPherson

INTRODUCTION

In the year 2002–2003, the City of Glasgow in Scotland (hereafter referred to as the City) introduced a Winter Festival that sought to agglomerate many existing cultural events and festivities under one banner, while also seeking to introduce new events to the calendar. The dual aims of this programme were to extend the tourist season by creating an 'artificial' winter festival and simultaneously providing entertainment for the local population. Other Scottish cities (see, for example, Foley & McPherson, 2004) hold 'Hogmanay' type festivities (which last up to six days after Boxing Day) during the winter season. What was unusual about the initiative in Glasgow was that organisers sought to extend their festivities in two directions simultaneously – across both time and cultures – by seeking to move beyond a celebration of Christmas as the only public festivity apparent in the City centre and branding the resulting entity as a single unified concept, namely 'Glasgow's Winter Festival'. The festival itself comprised of a wide range of events, starting with the political celebration of Guy Fawkes Night, and also including 'Glasgow Lights On' (illumination of the Christmas lights), 'Glasgow on Ice' (an open-air ice rink and fairground in the City centre), an Autumn/Winter shopping promotion, a further fairground in another City centre area, a continental market, 'Glasgow's Hogmanay', and ending with the 'Celtic Connections' folk music festival (all of these are subsequently referred to as 'the events'). Through this programme, the City sought to provide events that were accessible to all members of the culturally diverse population. It should be added that the time period covered by the Winter Festival also encompassed annual religious festivities of considerable significance to the City's Christian, Muslim, Hindu and Buddhist populations.

The aim of this article is to explore the relationship between the two policy objectives associated with the Winter Festival (the Festival); that of developing tourism and economic regeneration and that of meeting the cultural needs and improving the quality of life of the City's population. Superficially, at least, the branding of these events under one banner had the dual purpose of generating visitor income, while simultaneously serving the common 'needs' of citizens for identity and focus at this time of year. However, it is argued that it is simplistic to attribute the overall approach taken as ensuring the inclusion of diverse religions and cultural groups present in the City and thus securing engagement of communities within City Government and vice versa. Nor, it is believed, is it sufficient to label the Festival as developing an identity for Glasgow, and its citizens, as multi-culturally inclusive, tolerant and enterprising? It is also insufficient to see the festival solely as being emblematic of the

urban economic regeneration policies that have galvanised the retail and hospitality environments in the City centre and have led to the establishment of lifestyle-led and globally connected business opportunities in the services sector (e.g. Sykes and Roberts, 2000). While the article focuses specifically on the first official Winter Festival, it also provides an overview of the changes that have occurred since and the effects that these have had.

The purpose of this investigation was to examine whether such economic and community objectives can simultaneously be served through the cultural vehicle of a Winter Festival. This is done by comparing and contrasting the policy objectives set for the Festival with the actual outcomes. In considering the Festival, the concept of 'leadership for the common good', as expounded by Bryson and Crosby (1992) and Crosby and Bryson (2002), has been utilised, because it enables policy approaches to the (often contradictory) social, cultural and economic considerations of local government outlined above to be reviewed and contextualised simultaneously, even if they are not so easily reconciled in practice. This conception of the common good has been explored around policy issues from within the City Council's own leadership, via its objectives of inclusion, diversity and touristic development. Ostensibly, the Festival seemed to be serving both of these interests but a fuller investigation led to the questioning of whose economies were being developed, whose culture was being represented and which communities were being served. For example, there was little doubt that the tourist community (both consumers and suppliers) was being served with the promotion of the Festival and of Glasgow, generally, which was represented as a 'vibrant' place to visit in winter. Thus, the Council's objective of economic development was served, in part, at least. However, to what extent this met the Council's objective of 'making the City a more vibrant place at holiday periods for the citizens of Glasgow' (Glasgow City Council, 2002a) was less obvious. Probably, well-heeled residents of the City centre experienced a greater 'vibrancy' to their immediate living environments (whether welcome or not), but whether the Festival actually engaged more geographically and culturally 'peripheral' citizens and communities of Glasgow, or added to the social and cultural regeneration of the City, is worthy of a more detailed analysis.

'THE COMMON GOOD'

In attempting to analyse what lies behind the Glasgow's Winter Festival and the City's policies for leadership in events such as this, it was necessary to explore the Council's policy objectives and its expected outcomes from such an event. These policy objectives and their outcomes are analysed using Bryson and Crosby's (1992) idea of 'leadership for the common good'. These authors present a global view on the social and political world problems of the type that they believe can be solved through leadership and policy change aimed at achieving 'the common good'. This concept can be defined as a specific good or action that will be of benefit to most (if not all) members of a given community and is tantamount with the idea of general welfare, liberal democracy and benevolent collectivism. They argue that, in modern contexts, no one individual or agency holds sufficient power to lead and change but that, through 'shared power groups', public policy problems or issues can be better understood and resolved. Indeed, it is the case that most Local Authorities in Scotland work in partnership with other bodies to develop and provide culture (Cultural Commission, 2005). In the case of the City of Glasgow, this can specifically be seen in the recent introduction of Local Community

Planning Partnerships that encourage the public sector, voluntary organisations and community groups to work together to plan and deliver public services (Glasgow Community Planning Partnership, 2006). Glasgow's Winter Festival can be seen to address two such public policy issues – issues relating to social inclusion through events and festivals and those relating to economic growth.

For example, in the case of Glasgow's Winter Festival evidence of a cultural strategy that attempts to benefit the common good of the citizens of the City can be seen in socially inclusive policies, which attempt to benefit all citizens and races as well as allowing for an extension of well-established (and now globalised) Scottish festivity beyond Hogmanay. Furthermore, the Festival can be justified on economic grounds, in that it promotes the City's image globally, or at least nationally, thus attracting the visitors who spend money in the City and creating jobs for citizens. Moreover, there are considerations such as the promotion of City image to citizens themselves and securing the spending of these citizens within the urban centre rather than in peripheral economic entities outwith the administrative boundaries of Glasgow. Hughes (1999) sees this type of governmental behaviour as, simply, a new entrepreneurialism of the 1990s replacing the managerialism of the 1980s. He suggests that City management specialists (i.e. Council leaders) can both serve the economic priorities and simultaneously be culturally affirming through the development of such festivals and events in specific time periods and cycles (Hughes, 1999). In this model, City governments use culture as an agent of, and for, change, although the cultural product actually being produced may not be that which is asserted in the events themselves. Notwithstanding that important consideration, visitors and citizens are consuming an experience that, as Hughes (1999) argues, is culturally affirming and economically satisfying for the City managers. Thus, Glasgow has embraced the concept of culture as a key tool in helping to create and sustain regeneration, as highlighted by the transformation of Glasgow from a city that was once renowned for heavy drinking and violence to one of the UK's most cosmopolitan and cultural cities (Judd, 2003) since its status as European City of Culture in 1990. The City has sought to embrace the concept of cultural planning, whereby a culturally sensitive approach to urban planning and policy is adopted (Ghilardi, 2005). However, when there are competing (and, possibly, contradictory) common goods of both visitor spend and cultural diversity (as is the case in the City centre), it is reasonable to consider whose common good is achieved by the policies adopted. City image making represents a struggle for power, by elected members, Council departments and other stakeholders such as agencies promoting tourism or communities themselves (Lukes, 2002), and so, attendant political agendas merit specific consideration.

These issues, in turn, raise questions about whether, and how, Glasgow's Winter Festival has added to the social and cultural regeneration of the City or whether, like some other festivals (Evans, 2003), it is simply a commodification of a generic cultural experience for mass consumption. The lack of specific cultural distinctiveness in the naming of Glasgow's Winter Festival (other than a tenuous association with Pagan ceremonies of the seasons) and the events of which it consists, due to the absence of an artistic, music or religious connotation, suggests that it simply offers a vehicle designed to promote consumption among a mass audience that may produce economic benefits to the City and, as Hughes (1999) argues, is culturally affirming in that context alone. Perhaps it is simply the use of the City's available public spaces and its place identity (De Bres and Davis,

2001) that has provided the opportunity for the Festival to take place. Beyond the type of (rather tired) civic pride enlisted by the annual presence of colourful illuminations across parts of the City centre, there does not appear to be a public celebration of a specific cultural form connected to the City's past, present or future taking place in the festival.

Indeed, it is focused more around a set time frame rather than a particular celebration. At best, the main focal point may be Hogmanay revelry. Historically, in Scotland before the midpoint of the 20th century, that would have been a private family celebration – rather than a public spectacle as pioneered in Edinburgh (Foley and McPherson, 2004). But, Glasgow policy makers have sought to establish a festival over a winter period that moves beyond the spectacle of Hogmanay and have stated in their policies that they wish to 'enrich the quality of life of Glasgow's citizens by providing accessible, attractive and exciting cultural and leisure services' (Glasgow City Council, 2003, p. 1) while satisfying the Council's aspirational leadership in generating income from its own outdoor City spaces and thereby adding to continued regeneration. Pacione (1995) refers to Glasgow as continually pursuing initiatives that were designed to regenerate the City's economic base and specifically refers to Glasgow's Year as European City of Culture in 1990 as 'more to do with power-politics than culture' (Pacione, 1995, p. 250). Arguably, the Year of Culture was a success for Glasgow's City centre, and for tourism development, rather than for outlying communities or minority groups and those whose cultural capital fell short of the baselines necessary to participate actively in this new consumer phenomenon that had taken over the City (Wishart, 1991) – arguments that still retain contemporary currency (Khan, 2003). Indeed, Beckett (2003) states that, 'in truth, using culture to revitalise cities is an uncertain business' and Garcia (2003), in the same vein, comments that the longer term benefits to cities from these types of accolades are hard to measure, especially benefits for the residents. It is, therefore, reasonable to consider the extension of Glasgow's cultural policy to encompass the quality of life of its residents and reflect upon whether the City can reinvent its image through cultural regeneration (Sykes and Roberts, 2000).

Given that the Winter Festival is offered at a time of year that has become associated with a party atmosphere, celebration and the consumption of alcohol among some of the City's population, there are further, possible, public order-related interpretations of the Council's actions as a way to exert control over crowds gathered in central public spaces and to ensure that they are gathered in specific places where they can be monitored and controlled. Such leadership in regulation and prevention is also conceivable within a framework of the concept of 'common good', especially in areas associated with what the UK and Scottish Governments term 'anti-social behaviour'. The impact of these behaviours upon urban residents and businesses alike, and the image that representations of 'anti-social behaviour' project of the City, can be profound in a situation where it has struggled (and spent) to re-brand itself away from the 'No Mean City' (McArthur and Kingsley Long, 1984) of razor gangs, violence (see also Boyle, 1977) and sectarianism (Murray, 2000) to a paragon of post-Fordist economic restructuring through visitation, conventions and exhibitions, cultural populism, 'style' in retail merchandise offerings, attractive clubs and bars that are populated throughout the week (i.e. not only at weekends), a large student population and a reputation for multi-culturalism both among its citizens and in its public policies and practice. In this respect, Ravenscroft and Matteucci (2003, pp. 2–3) have suggested that 'festivals

are essentially internal political devices, used for the purposes of grounding cultural meanings within current regulatory frameworks', thus reinforcing the arguments of Hughes (1999) in the context of publicly supported Hogmanay festivals being used more as a means of social control than of promoting cultural policy or any widely held concept of "common good".

Jordan (1989) suggests that the concept of the 'common good' is notoriously difficult to isolate and even more difficult to deliver because of the inevitability of unforeseen circumstances intervening to frustrate intentions. He argues that only in Utopian situations is there a possibility that everyone does what is good for all, mainly because one person's idea of what is 'good for all' differs from another, culturally, politically and temporally. So operationalising conceptions of the 'common good' becomes a form of bargaining about provision by leaders and empowered   individuals   and   groups. However, Jordan (1989) also suggests that if policies by central or local government are to meet their targets, then communities' sense of collectiveness requires an active citizenship. Although this places responsibility onto individuals, Jordan (1989) is clear that they must be guided by political policy that is demonstrably inclusive and treats citizens as stakeholders. Bryson and Crosby (1992) discuss their concept of the 'common good' as involving collective action for public benefit, and they too agree that everyone involved needs to be at the level of stakeholders. They suggest that it is not possible to solve public problems or provide particular services in isolation from citizens, communities and other collectivities. In their conception, leaders need to utilise networks and organisational understanding as well as mobilise others, both political and ethical, to reach genuine collective achievement (Bryson and Crosby, 1992). Both of these sets of authors argue that, for a 'common good' to be achieved, some form of 'bargaining' or 'battle' must take place between those vying for power. Essentially, this is where the contest of power occurs, and those who are in power must find a way to work together to share power. In later work, Crosby and Bryson (2002) argue that this does not mean sharing power informally but that actual leadership must be shared meaningfully if public policy is to benefit the 'common good'.

Thus, the initiation and promotion of Glasgow's Winter Festival may be examined in two opposing ways. First, it may simply be an example of 'top down' political pragmatism to achieve the benefits of economic development, civic boosterism and behavioural control through commodifying some cultural contexts towards those ends. Such an interpretation could demonstrably serve some form of benefit embedded in neo-Liberal conceptions of the reasons for state intervention in social life, as being necessary only to secure the rule of law and of 'third way' ideas of intervention to develop the economic base of marginal businesses and, thereby, secure employment for citizens (Giddens, 1998). Furthermore, this analysis suggests (in part, at least) a political environment that could be seen to reflect the Paul Veyne's (1976) concept of the 'bread and circuses' basic (welfare) state, where the provision of human needs is combined with publicly held spectacle as a means of securing support for the policies and continuation of ruling elites. Alternatively, the Festival could be seen as being embedded in a collective negotiation and eventual consensus and shared conception of the role and nature of culture in meeting the aspirations of active citizens for their communities in the winter period of the year. In that case, it serves as a cultural festivity, satisfying the cultural policy agenda of providing culture for its citizens while, secondarily, generating economic benefit for the City (or retaining it within the City) through retail expenditures.

The discussion will now turn to which of these two analyses is dominant within the Council's approach, both in the perceptions of key stakeholders and in the practice of the festival itself.

RESEARCH METHODOLOGY

The research process started from a conceptual framework of cultural policy and leadership and provided the basis on which to conduct the fieldwork. An analysis of Council minutes provided evidence on the decision-making process in relation to the festival, and a combination of interviews with key stakeholders and an analysis of the policy documents that impact upon Glasgow's Winter Festival formed the fieldwork elements of the research. Interviews were carried out with the Lord Provost, with the Events Manager in the Land Services Department and with the Convenor of the Development and Regeneration Services Committee, in order to gain their views and perspectives on the role of events and festivities in the City and, more specifically, on the Winter Festival, the origins of its aims and whether they were being fulfilled. A number of formal Council papers and policy documents that were seen to be relevant to the Festival were analysed. The full Council Minutes from 2000–2006A were analysed in order to give a context to the initial development of the festival and to how and why it has changed since its initiation.

This article now goes on to review approaches to leadership of Glasgow's Winter Festival within the dual purposes outlined above. The data, when taken with the conceptual material presented earlier, point to three main areas of analytical possibility. First, issues concerning the role of public resources (human, land and so on) in the development of the Festival product and ancillary opportunities are raised. Second, the plurality of stakeholders in and around the Festival leads to further questions of the way in which leadership for the common good can be achieved. Last, and as a direct result of the previous two points, the role of culture within a context of consumption needs to be problematised.

FESTIVITY, CULTURE AND LEADERSHIP

Much of the success of Glasgow's economic and cultural regeneration in recent years can be attributed to the hosting and staging of events, which now forms an important part of both the City Council's overall income generation and its specific Cultural and Leisure Policies. Indeed, the City now promotes its cultural presence through a series of events that can attract equally the business investment and media attention as a means of securing their success. The Council produced an event plan as part of their Best Value Review (BVR – Local and Unitary Authorities in the UK have undertaken reviews of their service provision through a process called Best Value Review. This is designed to ensure that each Department's area of work is delivered to the Best Value for the money available. The process aimed to make service delivery more efficient and effective) conducted in 2001–2002, which highlighted the

> economic, tourism and social benefits derived from a well planned, organised and vibrant cultural, civic and sporting events programme for the City... and social, cultural and sporting benefits and the impact these have on widening access and participation (Glasgow City Council, 2002b).

Thus, two objectives of the Council's events strategy are made clear through this committee paper: to secure both economic and social benefits to the people of Glasgow. All other things being equal, there is no reason why both of these objectives should not be achievable

simultaneously, as reflected in the dual aims of the Cultural and Leisure Services Department (among others) to increase economic development and enrich the quality of life of Glasgow's citizens (Glasgow City Council, 2003). Indeed, the Lord Provost suggested that one type of regeneration engenders the other, stating

> the balance is very fine but, to put it in a nutshell, you will not economically enrich this City if you don't have the people socially involved and included. And, on the other hand, we won't be able to regenerate properly, socially, unless you've got some measure of economic regeneration (Interview with Lord Provost, 13 August 2003).

However, potential issues of achievement arise when those with responsibility for each are in separate Departments governed by different Council Committees, possibly with different political agendas. Indeed, a level of conflict could easily occur given that, on the one hand, the Land Services Department was responsible for bringing outdoor events to the City, generating income from the City's property holdings and for ensuring that the infrastructure was in place to hold such events, whereas, on the other hand, the Cultural and Leisure Services Department was responsible for the programming of festivals, for example, securing bands to play at Hogmanay. On that subject, one officer opined that the two Departments responsible for the Festival managed to work together, but that the internal cultures were not necessarily congruent, with both of the Departments having to work within their own events management framework to ensure that both their own Department objectives were met as well as those of the overarching Council (Interview with Events Operations Manager, August 2003). Moreover, this was complicated by the involvement of several other Council Committees such as Policy and Resources and Development and Regeneration.

During this time period, there was no one over-arching leader or group responsible for running the events, whether cultural or sporting in the City's public government, with respondents confirming that while working together, they also had to share leadership in the organising and running of the Festival. It should, however, be pointed out that this situation has now changed, with the Council currently having a dedicated events team.

The committee paper on events policy also stated that 'at a community level, community organised events can be a powerful tool in stimulating community involvement, pride and a sense of identity' (Glasgow City Council, 2002b). This seems to suggest that community involvement, and thus community gain, is best achieved via small-scale events that are organised by 'the community' themselves. It is difficult to reconcile this with the Winter Festival, which was intended to help make the City a vibrant place during the holiday periods, as reflected in its twin objectives of civic celebration and economic regeneration (Development and Regeneration Services, 2001). It is tempting to speculate that this was aimed at particular types of citizens, specifically those who had a particular set of cultural values and who could participate in the City centre festivities according to standards of economic and participatory behaviour consistent with the strategies of economic 'freedom' and disciplined 'control' adopted. Taken at its worst, one local newspaper's arts critic believed that offerings were exclusive to those with the cultural capital to be 'in the know' of how to get tickets for those parts of the Festival that required them and with the economic capital to consume these and other retail opportunities. Waterman (1998) refers to the upsurge in arts or cultural festivals as 'carnivals for the elites' and suggests that cultural festivals reflect the social order of particular social groupings. Specifically this refers to Zukin's (1995) contention that, instead of culture just 'being', it is now

used as 'an agent of change, no longer solely a reflection of material civilisation but rather an active force, manipulating images not only into saleable commodities but also to form the basis of tourist and real-estate markets and to inform visitors of collective identity' (Zukin, 1995, p. 113; Waterman, 1998, p. 55).

An associated problem appears to be that the Festival was not easily associated with one cultural form; indeed, the very core of the identity of the event is hard to encapsulate. For example, the Celtic Connections music festival (January 2003) was part of the overall Winter Festival, but so too was the fireworks extravaganza (5 November 2002) and the Hogmanay celebrations (31 December 2002), each using a different style of music and using it in different ways. So, the identity of the Festival was not entirely manifest or transparent. At best, it seems that this could be construed as a form of mass entertainment over a prolonged period that was presented as a portmanteau, generic cultural festival. However, it should be noted that the Festival was developed and built upon a number of successful initiatives that had taken place in the years preceding the first 'official' Winter Festival. In particular, it built upon the success of a temporary ice rink located in the City centre which, in an independent evaluation in 2000, was found to achieve its targets of both attracting shoppers and providing a civic celebration for the festive season (Development and Regeneration Services, 2001). The merging of various celebrations by policy makers to form an 'extended' festival can be seen to be based not only on a social/political need to promote local identity and provide entertainment for its citizens, but also to meet an economic/political imperative to derive a regional identity sufficient to support tourist visitation.

Glasgow is the third most visited city in the UK, attracting 3.2 million visitors, who generated £700 million for the local economy in 2003 (www.seeglasgow.com). Thus, the spending of public funds on local festivals can be presented as an attempt to invest in the quality of life of all citizens, while, simultaneously, tying activity to the strategic objectives of the Council to generate income from its estate and the infrastructure to enable revenue generation in local businesses. Glasgow's Lord Provost is quite clear that culture and leisure policy can be used to regenerate the City, both in economic and social terms, and that festivals and events are the key parts of the City's policy. She argued that

> there will come a day when Council budgets collapse, as they did in 1996 and you need to put a really strong policy of regeneration and strategic policy for inclusion there so that on rainy days you have something to fall back on. Well, we had to be opportunistic and I have put this strategy in place... one of those strategies was to get opportunities for a year here or there (in terms of festivals and events) but also having festivals in sporting or cultural fields which would profile the City but also contribute to regeneration; Glasgow's Winter Festival is a microcosm of that (Interview with Lord Provost, 13th August, 2003).

Furthermore, a national dimension arising in the Council's draft report on strategic objectives and key actions in relation to events augments the local and regional elements cited above. This stated that one of these strategic objectives would be to

> capitalise on existing and potential benefits of hosting major events. Glasgow and all key partner agencies, both public and private, must challenge current perspectives and work towards creating a common vision and shared priorities which reflect the Council's key objectives and existing strategies (Glasgow City Council, 2002c).

This appears to link directly to the new national agency for events promotion in Scotland, EventScotland, given that its

Major Events Strategy (2003–2015) provides the objectives

> to deliver a viable portfolio of major events to attract visitors to Scotland, to enhance Scotland' international profile, to strengthen our sporting and cultural infrastructure and to maximise the economic, social and environmental benefits of events to all parts of the country (Scottish Executive, 2003, p. 1).

Turning to the participation of a plurality of stakeholders in the Festival, it could be argued that the approach of Council policy makers was to involve local community groups. Certainly there were various partnership bodies involved in the organisation and funding. The Central Area Forum (a business-led interest group), in particular, was involved in the development of the Festival and provided funding for a number of light installations in George Square (the epi-Centre of the City) during the Festival (Central Area Forum, 2002), while sponsorship was also received from the Clydesdale Bank for the Fireworks display on Guy Fawkes Night and from Radio Clyde (a local Scottish independent radio station) for the Hogmanay event. However, there was little consultation in the way that specific community groups could/should and could not/should not be included in the Festival according to some respondents. One elected member thought that what he described as 'minority communities' were under-represented at the Winter Festival generally and at Hogmanay in particular. However, he also argued that he did not support the promotion of culturally specific events for different cultural communities and that, in the past,

> what has happened is that we expected (minority) communities to run their own festivals, with a lot of people throwing money at them, and this has sometimes led to conflict in the communities themselves. Some communities are better at doing things than others are, and there's a perception that perhaps they

are receiving more assistance' (Interview with Glasgow City Council Councillor, 28 July 2003).

He went on to highlight

just because a group of people are not very well articulated in terms of making presentations and applications, they would suffer for the following years because they would be unable to produce the level of clarity the Council were looking for (Interview with Glasgow City Council Councillor, 28 July 2003).

He suggested that this was why minority groups needed the Council to run festivals and events all year-round and not just for one particular group. He argued that, in this way, the Council can ensure that festivals are open to all and are as inclusive as possible. He went on to state that

I think we live in an era where we, for sometime, have lost our culture as a Scottish nation and we have to re-establish that culture, and that culture is going to be re-established by all the cultures that we have in Scotland today (Interview with Glasgow City Council Councillor, 28 July 2003).

From his perspective, events such as the Winter Festival helped to promote a Scottish life that is multi-cultural, that there was a need to re-establish these values nationally and that this was a way of bringing culturally diverse communities together. In terms of the role of the Council, he suggested that, formerly, people organised these events for themselves albeit, not on this scale, but that the erosion of communities and their collective identities had led to the loss of any meaningful cultural cohesion. In his view, this coherence was now provided by the Council in events such as the Winter Festival. He argued that 'these events are helpful because they do bring people out, and they do encourage communication between communities, but the council will have to do more to encourage specific groups in

communities' (Interview with Glasgow City Council Councillor, 28 July 2003)

Although it was recognised that the Winter Festival provided a vehicle for bringing together people from different communities within the City, the success of the Festival in doing so was questionable with one particular community in Glasgow – a community that had received a high profile during 2002–2003 – being publicly uncomfortable with the Winter Festival events. The local press reported that asylum seekers who had been housed in Glasgow had felt excluded because they did not get tickets or did not know how to get tickets for the main Hogmanay event. This was an opportunity lost to recognise all members of Glasgow's diverse community but, also, it highlights the difficulties that the local authorities face in dealing with issues of social capital and ensuring that cultural strategies are accessible to a range of groups. The wider issues raised by this criticism, however, led to questions about who possessed sufficient cultural capital to participate as an included citizen and raised the issue of whether there was a question of accessibility (social constraints and psychological inhibitors) rather than just access (physical constraints; economic and spatial inhibitors). One Councillor did argue, however, that

there will always be a segment of the community who will be unaware of what's happening in the City, you will never, ever, be in a position to be able to inform every single individual member of the community about what's happening (Interview with Glasgow City Council Councillor, 28th July 2003).

He also asserted that the City had worked hard to ensure that all sections were represented but added that 'sometimes it is just first come first served in a democracy and that's the way it works' (ibid). Thus, the difficulty with city-wide social inclusion policies can be that, for an event like the

Winter Festival, which is presented and promoted as generic and as an opportunity open to all, the lack of evaluation evidence makes it impossible to judge whether that outcome of inclusion is met. Moreover, if that aim is not met, it remains unclear what factors may have contributed to failure and which are susceptible to manipulation to encourage future inclusion.

In line with the remit to provide events, which integrated the promotion of Glasgow with the City's social justice agenda (Glasgow City Council, 2002a) formal attempts, were made to create an opportunity for access to the Winter Festival. The vision of social inclusion and 'common good' for the people of Glasgow was observed through the provision of subsidised tickets and the organisation of school trips for children from outlying Social Inclusion Partnerships (SIP's – these were a governmental initiative whose aim was to work with specific communities who were deemed to be economically and socially deprived). These have now been replaced by Community Planning Partnerships (CCP – government agencies whose aim is to bring together the key public, private, community and voluntary representatives with the aim of delivering better, more joined up public services within the City across the whole of the City, thus representing all of the City's communities) areas, such as those in Castlemilk and Easterhouse (two of the most economically and socially deprived areas in the City, indeed, in Scotland) to participate in activities such as the City centre ice rink and associated carnival. This enabled some people from what are termed 'minority and excluded communities' to participate in one aspect of the Festival (i.e. commercially provided and priced service in the market place). However, many of the same populations appeared to be excluded from participation in collective opportunities such as the Fireworks, 'Glasgow Lights On' or Hogmanay events, which were free to attend but required family, or other informal support

networks, to enable a child from an outlying area to attend.

In attempting to serve both economic and social regeneration agendas, the Winter Festival sought to bring money and culture together, as was achieved when the City gained the title of City of Culture in 1990. Wishart (1991) suggests that this is acceptable and argues that the hype around the Year of Culture and the promotion of arts and culture through events led the City to a 'civic rebirth' and also suggested that culture was a vital tool in the urban regeneration that followed. But, to take Zukin's (1995) argument, it seems possible that the culture celebrated is being lost at the expense of it being used as an 'agent for change'. Whether or not this is the case with Glasgow's Winter Festival is up for debate, with one respondent suggesting that, although the Council are keen to promote tourism, they are not willing to allow certain types of Scottish and Glaswegian culture (which, by their nature, are exclusionary) to be eroded.

Equally, it could be added that these same local cultural forms of expression through festivity and remembrance are being replaced with global expectations of consumption, spectacle and universality and that public policy of the type pursued by Glasgow via its events leads to complicity in this change towards a culture of commoditisation. More likely, what was being witnessed was another re-invention of Glasgow as a place of festivals and events – it is too easy to romanticise long-standing festivities as being culturally 'pure' and untainted by commercialism or 'trade'. In fact, successful festivities have long attracted ancillary 'side-shows' where market offerings sit alongside identity, meaning and cohesion. It could be argued that such a relationship is the very essence of the common-good, reaffirming, as it does, the necessities of economic stimulation and the imperative of collective identity as (mutually supportive) partners in dynamic communities.

It seems that Council spending on events is a key to public policy and that it expects to develop a new urban entrepreneurialism and serve a duality of social purposes. The Lord Provost stated that

> culture is very much an agent of change and that the key word is sustainability... You must take people with you, so you must have things that impinge upon their daily lives and on the community as well as being big blockbusting things; you must make sure that there's a cultural policy strategy in place and you need to be working after events (Interview with the Lord Provost, 13th August 2003).

This appears to reinforce the argument that culture can be used as an agent for change and that, in Glasgow at least, the common good of citizens and tourists are, in policy terms anyway, seen as going hand in hand. Specifically, in relation to Glasgow's Winter Festival, the Lord Provost said that this idea arose from Nuremberg (with which Glasgow is twinned), where the idea was to link the economic regeneration of getting people into the City shopping throughout the holiday period, with amusement and entertainment. She argued that it was also,

> about making it very much for the people of the whole city to come in, it was a huge focal point of discussion in the Labour group, about encouraging folk to come to the ice rink; that whilst you're fronting this as 'come to Glasgow and enjoy the winter in Glasgow', we're also making conscious efforts to ensure that school trips get there, that those tickets are subsidised, and that people are encouraged to make sure that they feel that this is part of their City, and that they're not being put to one side while we get to the tourists (Interview with Lord Provost, 13 August 2003).

This, on the surface at least, appears to reinforce a holistic policy position of serving a common good that is actually both economically and socially desirable. Nevertheless, these concepts require to be continually interrogated if they are to be asserted with confidence in terms of the common good.

CONCLUSIONS

It would appear that, despite some argument to the contrary, Glasgow's public policy in terms of the first 'official' Winter Festival is more conducive to income generation, City entrepreneurialism and post-modern spectacle rather than to cultural participation and reach among all citizens. But, arguably leadership in the former leads to the latter. The common good that is served can arise through increased income to tourism via local businesses to local people through concomitant effects of job generation, spatial aesthetics and, possibly, community harmony. Nevertheless, while job generation is often promoted as a benefit and outcome of attracting large numbers to festivals, Garcia (2003) reminds us that these jobs are often temporary, low-level service jobs. Thus, festivals that are time and place specific are not likely to be the answer to long-term urban regeneration, nor does it seem that they fully satisfy the local community's cultural needs (Evans, 2003) – although that may depend upon how the community views itself! The Winter Festival of 2002–2003, for some, was a lost opportunity for Glasgow to demonstrate a cultural event that could reflect its multi-cultural values and transcend tensions about immigration that have been part of the City's identity for hundreds of years. In part, this perceived that the lost potential was due to the lack of representation of these diverse cultures in the Festival programme. However, on a more positive note, the generic cultural experience that was presented through Glasgow's Winter Festival in 2002–2003, and especially in the three Winter Festival programmes since, do appear to serve a common good, although not, as Bryson and Crosby (1992) had suggested, as a cohesive policy approach but, rather, as a festival that served a common good by being a culturally generic experience for many and one which was obviously there to be happily (and consciously) bought and consumed. Glasgow has no reason to apologise for what it can do well and should celebrate an unashamedly populist approach to cultural consumption as a key means of contrasting itself with other competitors in Scotland and beyond. Within this (superficially) simplistic observation lies the implication of sharing power in leadership of cultural policy. Sharing implies the necessity of compromise, a concept that sits uneasily with issues of authenticity in culture. Such a compromise is likely to be most easily achieved in the province of the 'popular' or the 'generic' rather than in the 'local' or the 'specific'.

Since the 2002–2003 Winter Festival, Glasgow's festivals have served to highlight the growing trend in UK cities to see festivals as commodities. Indeed, with the establishment of the EventScotland agency, it seems that selling Scotland and its culture as a commodity is becoming more important for cities, particularly in terms of their cultural and regeneration strategies. It could be suggested that the policies out of which the first Winter Festival was born were somewhat utopian – i.e. that the Festival could provide a balance of benefits for tourists, business and the local population. This is not to say that festivals and events cannot serve these multi-faceted purposes, but it is clear that dynamics of the relationship is more complex than generally presented and the desirability of simultaneous achievement of these is, at least, questionable as a liberal democratic shibboleth.

CODA

Glasgow's Winter Festival has developed significantly – many of the events have had to find new and additional sources of funding,

for example the Central Area Forum no longer funds the Christmas lighting in George Square and new sponsors have been found for the Fireworks display, which has led the Festival to become increasingly commercialised. Furthermore, the Festival continues to offer a generic cultural experience with the only real major changes involving the provider of the ice rink and the moving of the Continental Market, which now takes place for several weeks in the run up to Christmas rather than for just a few days. It would appear that Glasgow City Council has recognised that the Festival cannot provide for economic and social regeneration in equal proportions, with a greater emphasis now being placed on the economic aspects of the Festival. This is, however, not to say that the social agenda has been forgotten, with the Festival continuing to organise and subsidise the use of the 'Glasgow on Ice' facilities for school children from (so-called) deprived areas. Moreover, a George Square Festive Group was set up in 2005 to unify plans for the use of George Square over the coming festive seasons (Central Area Committee, 2005), which, as well as combining the input from several Council Departments with responsibility for organising the Festive events, will allow the people of Glasgow to have a greater say in how the Festival is organised (in George Square at least).

REFERENCES

Beckett, A. (2003) Can culture save us? *The Guardian*, 2 June.

Boyle, J. (1977) *A Sense of Freedom*, London, Pan.

Bryson, J. M and Crosby, B. C (1992) *Leadership for the Common Good, Tackling Public Problems in a Shared-Power World*, San Francisco, Jossey-Bass.

Central Area Forum (2002) *Council Minutes 9th December 2002*, Glasgow, Glasgow City Council.

Central Area Committee (2005) *Council Minutes 17th October 2005*, Glasgow, Glasgow City Council.

Cultural Commission (2005) *Cultural Commission Final Report*, Edinburgh, Scottish Executive.

Crosby, B. C and Bryson, J. M (2002) *Conceptions of the Common Good*, Paper presented at University of Strathclyde, October 18.

De Bres, K. and Davis, J. (2001) Celebrating group and place identity: A case study of a New Regional Festival, *Tourism Geographies*, **3**(3), 326–337.

Development and Regeneration Services (2001) *Council Minutes 22nd March 2001*, Glasgow, Glasgow City Council.

Evans, G. (2003) Hard-Branding the cultural city – from Prado to Prada, *International Journal of Urban and Regional Research*, **27**(2), 417–440.

Foley, M. and McPherson, G. (2004) Edinburgh's Hogmanay: In the Society of the spectacle, *Journal of Hospitality and Tourism*, **2**(2), 29–42.

Garcia, B. (2003) In A. Beckett, Can Culture Save Us? *The Guardian*, 2 June.

Giddens, A. (1998) *The Third Way: The Renewal of Social Democracy*, London, Polity Press Blackwell Publishers.

Ghilardi, L. (2005) *Culture at the Centre, Cultural Planning: A Strategic Approach to Successful and Sustainable Community-Based Regeneration in Scotland*, Edinburgh, National Cultural Planning Steering Group.

Glasgow City Council (2002a) *Cultural and Leisure Services Annual Review 2001–2002*, Glasgow, Glasgow City Council.

Glasgow City Council (2002b) *Report produced on a committee paper on benefits of events arising from the BVSR of Arts and Cultural Events and Draft Review of Sport and Recreation in Respect of Events*, Glasgow, Glasgow City Council.

Glasgow City Council (2002c) *Cultural and Leisure Services: Events Key Issue Working Group Draft Report on Strategic Objectives and Key Actions for 2002/2003*, Glasgow, Glasgow City Council.

Glasgow City Council (2003) *Cultural and Leisure Services Annual Review 2002–2003*, Glasgow, Glasgow City Council.

Glasgow Community Planning Partnership (2006) *Our Vision for Glasgow: Community Plan 2005–2010*, Glasgow, Glasgow Community Planning Partnership.

Hughes, G. (1999) Urban revitalization: The use of festive time strategies, *Leisure Studies*, **18**(2), 119–135.

Jordan, B. (1989) *The Common Good: Citizenship, Morality and Self-Interest*, Oxford, Blackwell.

Judd, D. (ed) (2003) *The Infrastructure of Play – Building the Tourist City*, New York, ME Sharpe Inc.

Khan, S. (2003) What Did Culture Ever Do for Us? *The Observer*, 8 June.

Lukes, S. (2002) Review article power and agency, *British Journal of Sociology*, **53**(3), 491–496.

McArthur, A. and Kingsley Long, H. (1984) *No Mean City*, London, Corgi Adult Publishers.

Murray, B. (2000) *The Old Firm*, London, John Donald Publishers.

Pacione, M. (1995) *Glasgow: The Socio-Spacial Development of the City*, Chichester, John Wiley and Sons.

Ravenscroft, N. and Matteucci, X. (2003) The festival as carnivalesque: social governance and control at Pamplona's San Fermin Fiesta, *Tourism, Culture and Communication*, **4**(1), 1–15.

Scottish Executive (2003) *Scotland's Major Events Strategy 2003–2015: 'Competing on an International Stage,'* Edinburgh, Scottish Executive.

See Glasgow (2006) accessed at, http://seeglasgow.com, accessed 11 August 2006.

Sykes, H. and Roberts, P. (2000) *Urban Regeneration: A Handbook*, London, Sage.

Veyne, P. (1976) *Le pain et le cirque: Sociologie historique d'un pluralisme politique* (Paris: Seuil, 1976), abridged in English translation: *Bread and Circuses: Historical Sociology and Political Pluralism*, tr. Brian Pearce, (1990) intro. Oswyn Murray, London, Penguin Press.

Waterman, S. (1998) Carnivals for Elites? The Cultural Politics of Arts Festivals, *Progress in Human Geography*, **22**(1), 54–74.

Wishart, R. (1991) Fashioning the future: Glasgow, in: M. Fisher and U. Owen (eds) *Whose Cities?* London, Penguin.

Zukin, S. (1995) *The Cultures of Cities*, London, Blackwell.

Interview with Convenor of Development and Regeneration Services Committee, 28th July 2003.

Interview with Lord Provost, 13th August 2003.

Interviews with Land Services Department Events Operations Manager, August 2003 and July 2005.

Mentoring volunteer festival managers: Evaluation of a pilot scheme in regional Australia

Anne-Marie Hede and Ruth Rentschler

INTRODUCTION

Festivals have a number of functions in society. They build community pride (Wood, 2002), celebrate important occasions and anniversaries (Clark, 1981), and provide opportunities for the exchange of information or goods and services within, and between, communities (Ryan, 1995). For individuals, festivals encourage community members to reflect upon and continue their culture's history and traditions. For example, Sofield and Li (1998, p. 277) observed that the 800-year-old Chinese Chrysanthemum Festival perpetuates and venerates traditional culture and is 'an expression of both thanks for past prosperity and hopes for future wealth'. As a type of special event (Jago, 1997), festivals also provide individuals with opportunities: to escape the routines of their everyday lives; for novel experiences; to satisfy their curiosity; and to be with people who are enjoying themselves (Formica and Murrmann, 1997;

Kerstetter and Mowrer, 1998; Saleh and Wood, 1998; Uysal et al., 1993).

Festivals are agents of cultural and social change. As such, their organisers are 'expected to act like [organisers of] other industries, developing professional management skills, performance indicators and other markers of a mature and ultimately self-sustaining enterprise' (Craik et al., 2003, p. 22). Many festivals, however, are managed by volunteers (Spring, 1995), and while volunteers possess a range of skills, these are not always honed for the tasks associated with festival management. Both Regional Arts Victoria (RAV) and Arts Victoria (AV), the two main government instruments that support arts festivals across the State of Victoria in Australia, noticed this to be the case in the arts festival sector in regional Victoria.

While there are many courses in Victoria, and in Australia, aimed at training festival managers, most are delivered through the formal education system – and these are not always accessible for those volunteering their time to regional festivals. Furthermore, they are not always designed with the arts festival in mind. In response to these issues, RAV and AV developed a pilot mentoring program for arts festivals in regional Victoria, *Directions*. The premise of *Directions* is that if festivals are managed well, their outcomes will be enhanced for their various stakeholders. This paper is based on an evaluation of *Directions*.

The research undertaken for this paper aimed to identify the areas of management that festivals in regional areas perceive to be of importance to them, and determine whether mentoring was an effective method of skill enhancement in this context. The paper then provides a background to mentoring as a management tool generally, and more specifically in the arts sector. Following this, the Australian regional arts festival sector is described and an overview of *Directions* provided. The method employed for the evaluation of *Directions* is outlined, followed by an overview of the six festivals that participated in the pilot mentoring program. The results are then presented and synthesized, reflected upon, and discussed. Finally, recommendations are posed for the future of mentoring in this context and for further research on the topic.

While this study is practically focussed as it examines mentoring within the context of arts festivals, it also contributes to the literature on volunteers (cf. Geroy *et al.*, 2000; Jago and Deery, 2000; Lourdes, 1999; Whitaker, 2003; Williams, 2003). Given that the program was targeted at volunteers of arts festivals, the study contributes to the growing body of literature on volunteers within the context of festivals and events (cf. Davidson and Carlsen, 2002; Elstad, 1996, 2003; Pegg, 2002; Saleh and Wood, 1998; Slaughter, 2002; Solberg *et al.*, 2002; Williams *et al.*, 1995). To date, the literature on volunteers undertaken within the

context of festival and events has highlighted some interesting findings. As example, Solberg *et al.* (2002, p. 26) investigated the 'value' of volunteers at the 1999 World Ice Hockey Championship and found that a large number of the volunteers at that event 'received so-called psychological returns from [their participation]' and that 'more than 80% would do the job over again'. Furthermore, Elstad (2003) found that in the case of one jazz festival, over 30% of volunteers felt that they could have withdrawn their services at some stage during the course of their involvement in the festival. Noticeably, however, much of the research on volunteers within the context of festivals has focussed on volunteers at the operational level. Little interest, however, has been directed towards those volunteers that manage festivals and events. Hence, this paper addresses a gap in knowledge.

MENTORING AS A MANAGEMENT TOOL

RAV and AV decided that *Directions* would be mentor-based. As a management tool, mentoring has been traced to ancient Greece, although Tenner (2004) suggested that contemporary mentoring only emerged in the 1970s. Indeed, Tenner suggested that, today, mentoring is as much a career tool as is a curriculum vitae or a resume. DuBois and Neville (1997, p. 228) referred to Nettles' description of contemporary mentor programs as those that are seeking to establish 'a one to one relationship between [the mentor] and...the mentee in meeting academic, social, career or personal goals'.

Although there is ample literature on the topic of mentoring, particularly in education (cf. Booth, 1993; Edwards and Collison, 1995; Zanting *et al.*, 2003), and at-risk youth (cf. Freedman, 1988), much of it appears to lack a common understanding of what constitutes mentoring. Stone (1999) clarified some of the differences between mentoring and other related concepts, including coaching and counselling. He described coaching as the function of hiring, training and evaluating one's 'team'; counselling as the function of communicating with those on the team who are not performing well; but mentoring, which is directed at the best performers on the team, encompasses being a role model, coach, broker, and advocate. He further added that mentoring might involve sustaining the motivation of volunteers despite limited opportunities. There appears to be some overlap of these related concepts, but overall mentoring is a multi-dimensional activity with various foci that involves instructing, advising, providing feedback, evaluating and challenging the mentee (Zanting *et al.*, 2003).

DuBois and Neville (1997) noted that much of the contemporary research and literature on mentoring has focussed on the establishment of mentoring programs. Austin *et al.* (2002) described the evolution of one mentoring program, within the context of a university, and its transformation from an informal to a formal program through the process of review. In the case of that mentoring program, policies were eventually developed in relation to mentor co-ordination, activities, training for mentors and feedback. Other ways in which it was thought that the program might be improved included the use of online communication opportunities, social functions for participants, and the linking of the program into a formal subject for academic credit.

In terms of evaluation, Sipe (2002) noted that a number of studies have been undertaken to identify the reasons that mentor programs are successful. She noted that there is consensus in the literature that the 'key to creating effective mentoring relationships lies in the development of trust between two strangers' and that this critically relies on the approach of the mentor. Furthermore, Parra *et al.* (2002) highlighted

that previous research on mentoring has identified that 'more positive perceptions of program training and support should predict greater benefits and relationship continuation'.

MENTORING IN THE ARTS

In the arts, mentoring has been less goal-oriented than a traditional teaching relationship and success is measured by depth of knowledge and competency rather than content curriculum covered (Young and Perrewé, 2000). Mentoring in the arts focuses on creating and facilitating networks, enhancing pathways, and engaging in essential discourse. Mentor schemes are currently used for Indigenous, metropolitan and regional arts programs for this purpose by organisations such as the Australian Film Commission, Arts Victoria, the contemporary dance company Chunky Move, and the Alice Springs Youth Arts Group.

Mentoring relationships can offer arts practitioners opportunities to engage in experiences appropriate to their current abilities and their future potential. In addition to the observable skills required in the arts industry, individual success can depend on a range of complementary skills and knowledge not formally encountered by arts organisations, such as writing grant applications, networking with other organisations, flexible and creative approaches to engaging with the surrounding community (Hunter, 2005). The mentor relationship facilitates the enhancement of these types of skills by demystifying the processes and clarifying the pathways that lie before the arts practitioner.

ARTS FESTIVALS IN AUSTRALIA

The end of the harsh European winters provided the impetus for carnivals; the end of picking seasons signified an abundance of food for farming communities in the

USA – but these traditions were less meaningful for Australians. Hence, festivals emerged as a means in which to celebrate Australian traditions (Alomes, 1985). Arts festivals, in particular, have become an important part of Australian culture; most Australian capital cities boast an international arts festival, and increasingly regional towns are using arts festivals to showcase their region's talents and add cultural cachet to their town. Spring (1995) conducted the first survey of arts festivals in Australia for the Australia Council (the federal government's arts funding and advisory body). Until then, little was known about the contribution of festivals to the cultural industry.

Some 1300 festivals are now staged in Australia (ABS, 2002, 2003): a large proportion of them are arts festivals. In a population of near 20 million (Anon., 2005), this makes it a highly competitive industry. In the United Kingdom, there are 550 arts festivals (Allen and Shaw, 2004) servicing a population of 58.5 million (Anon., 2005). From a managerial perspective, the arts, including music, painting, sculpture, drama, opera and literature, can add creativity to festival programming – a successful artistic program can be an effective means of distinguishing one festival from another (Lade and Jackson, 2004). Producing and delivering the arts are, however, new territories for many festivals.

Australia tends to focus on its capital cities as the centres of activity. Melbourne, Sydney, and Adelaide, for example, have become the gateways to regional locations. Furthermore, there are vast differences regarding the pace of life and the services that are available in regional locations other than cities. Frost (2006) noted the declining employment opportunities in agriculture and forestry, the closure of many services and the exodus of young Australians to the cities, as factors that have impacted regional Australia. Consequently, regional Australia can be isolated – both geographically, and at times psychologically, from the larger capital cities. In the face of these attributes, however, the festival sector in Australia has become quite vibrant. Still, RAV and AV identified that there is a need to develop networks within the regional festival sector to enhance the outcomes of festivals in those areas.

DIRECTIONS AND MENTORING

While RAV and AV identified the need to develop networks and skills within the regional arts festival sector, they also recognised the challenges, including the tyranny of distance, time and resources, to provide opportunities for volunteer festival managers to develop their skills. Mentoring seemed to be a suitable way in which to overcome these challenges. *Directions* was a programme designed by RAV and AV to assist volunteer organisers of regional arts festivals to define and develop the artistic direction of their festivals by facilitating the establishment of mentees/mentors relationships.

A budget was set for *Directions* which allowed for six regional festival managers, who were also volunteers, to each be mentored by three experienced festival artistic directors. *Directions* was initiated by a call for expressions of interest, which detailed the requirements of participants, specifically that they would be required to attend a briefing session in Melbourne, Australia (which is the capital of Victoria), communicate with their mentors on a needs-basis over the following three months, and attend a seminar in a regional location coordinated by the RAV and the AV when the program would be discussed in a public forum. Successful applications were provided with a small budget to facilitate their participation in the program.

RAV and AV assessed the submissions and based on the quality of their submissions and their demonstrated need for mentoring selected six festivals to participate in the program. The final sample of festivals comprised six small festivals, including four music festivals, one theatre festival, and one craft festival. Spring (1995) observed that the festival sector in Australia comprises a few large festivals and a large number of small festivals, and that the most common type of festival is the music festival. The three mentors were selected by RAV and AV based on their experience as artistic directors in Victoria. Each mentor was assigned two mentees to work with for the program. Mentors and mentees were matched in respect of the mentors' experience, the festivals' art form, stage of festival development, and geographic location of both the mentees and the mentors. It was acknowledged that some mentors would be working with more than one mentee from each festival and this was deemed to be suitable, given the nature of festival management in regional locations. In most instances, the mentoring entailed two or three face-to-face meetings, and more frequent contact via telephone or email.

METHOD OF EVALUATING *DIRECTIONS*

As this was the first analysis of mentoring in regional arts festivals, there was no specific model on which to base the analysis. The *Five Stage Approach to Evaluation* developed by Woolf (2004) was used to guide the evaluation of *Directions*. Woolf's approach to evaluation has been used in the arts arena extensively (cf. Williams and Bowdin, 2005). The five stages to evaluation are planning, collecting evidence, assembling and interpreting, reflecting and moving forward, and reporting and sharing.

1. In terms of planning, the researchers and RAV developed an overall approach to the evaluation, that is, how long it would take, who would be involved, how the information would be collected, what questions would be asked of participants, and how the results would be collated and disseminated.

2. The evidence was collected via overt participant observation at the initial briefing session, via web searches of the festivals, and by undertaking semi-structured in-depth interviews at the conclusion of the program. Interview questions, which formed the basis of audio-taped telephone interviews, were developed taking into consideration the objectives of *Directions* and its method of delivery. This method of data collection was deemed to be appropriate, given the locations of the participants and the researchers. Questions included in the interviews related to managerial practices, artistic direction, expectations, and short-term and expected long-term outcomes of program participation. The same questions were asked of mentees and mentors. Each interview was approximately one hour in duration. In total nine interviews were undertaken – one with the key mentee from each of the festivals and one with each of the three mentors.

3. The authors then independently analysed the transcripts of the interviews for salient themes, focussing on two particular concepts: (i) the areas of management that were of concern to the festivals and (ii) whether the mentoring process was effective. The researchers then discussed their analyses, refined them, and assembled the data.

4. The researchers then reflected upon the findings and proposed ways in which the regional arts festivals could move forward using mentoring in the future.

5. The results were then disseminated at a public forum organised by RAV and AV, via FESTNET (a website for festivals

hosted by RAV) and a number of scholarly publications.

Brief profiles of the six participating festivals are now provided for readers. Following this, a synthesis of the information gained from the analysis is provided, which then form the basis of the discussion. Where direct quotes have been used, pseudonyms have been used to protect the identity of the mentors and mentees. However, it is acknowledged that individuals participating in the program may be identified nonetheless as the program has been discussed in public forums organised by RAV and AV previously.

PARTICIPATING FESTIVALS AND THEIR PROFILES

The six festivals that participated in *Directions* were The National Celtic Festival, Bruthen Blues and Arts Festival, The Stitched-up Textile Festival, Apollo Bay Music Festival, the Stratford on Avon Shakespeare Festival, and the Moss Vale Music Festival. While three of the festivals are situated in the Gippsland region, Gippsland covers a very large area; as such the festivals in this region are not in direct competition with each other. The festivals, while not chosen on the basis of the art they represented, provided a cross-section of the arts sector. The profiles of the festivals are provided in Table 1.

MANAGEMENT AREAS PERCEIVED AS BEING OF IMPORTANCE TO THE FESTIVALS

Seven key management areas were identified as being as of importance to the participating festivals.

Strategic Development

Strategic development was raised at the Bruthen Blues and Arts Festival, the National Celtic Festival, and the Moss Vale Music Festival. The themes that emerged under this heading were the development of a strategic business plan, developing strategies to broaden the festivals' appeal to markets, and the need to sharpen the appeal of the festivals through clearer artistic aims. The mentoring process enabled the mentees to examine past and present practices so that they could identify opportunities for future direction. The mentors encouraged the mentees to connect strategic planning and artistic direction more explicitly. For example, one Mentor, Catherine, encouraged the mentee for the Stitched-up Textile Festival to view a change of venue as an opportunity to recreate the opening event and as a way of getting people into the festival zone, and adding value to the festival through art.

Stakeholder Relationships

A number of participants raised the problems associated with isolation in regional locations, and the mentoring process specifically focussed on the developing relationships with stakeholders to address this problem. For the Moss Vale Festival, in South Gippsland, for example, networking and building links with other organisations, artists, and community groups in regional areas were determined as being invaluable to the festival's sustainability. It was also a way of enhancing interesting opportunities for program diversification. A number of relationships had already been forged by the festival, but these were not instantly apparent to the festival's mentee. Mentoring for artistic direction assisted in highlighting these relationships. For example, while in 2004 hay bales were used to buffer the sound of a very noisy generator as a result of involvement in *Directions* in 2005, they were used as part of an art installation. The hay bales emerged as a symbol of the Moss Vale region, the plight of the local farmers in recent years and a celebration of the recent harvests that the farmers were experiencing. Thus, *Directions* was instrumental in

Table 1 Participating Festivals and Profiles

Festival	Location	Theme	Duration	Activities	Feature
National Celtic Festival	Bellarine Peninsula	Celtic heritage	3 days	Street Performances Workshops	Australia's largest arts and cultural Celtic event.
Bruthen Blues and Arts Festival	East Gippsland	Blues music	2 days	Performances Activities	Biggest Little Blues Festival in Australia. 10 year celebrations in 2005.
The Stitched-up Textile Festival	North East Victoria.	Regional textiles	2 days	Workshops Displays Competitions	In operation since 1998. Change of venue in 2004 to Showground Pavilions.
Apollo Bay Music Festival.	Geelong Otway Region	Music of any genre	2 days	Performances Activities	Operating for 10 years. Well-established brand in the market place.
Stratford on Avon Shakespeare Festival	Gippsland	Shakespearian	10 days	Village fair Workshops Performances	First staged in 1991.
Moss Vale Music	South Gippsland	Folk music	2 days	Performances Activities	World music festival

identifying, accentuating, and articulating the connection between the festival and members of the local community.

Following on from this, the opening of the Commonwealth Games in Melbourne coincided with the 2006 Moss Vale Music Festival. As a direct result of *Directions*, the festival approached the local Shire as to how they could work together to harness opportunities arising from this international event. Moss Vale Music Festival's world music image became a feature of the Shire's Commonwealth Games event strategy. Further, the organisers of the Moss Vale Music Festival also worked closely with the Indigenous community to ensure that it was represented in the 2006 program, as Indigenous Australia was a key plank in the Commonwealth Games cultural program.

Audience development

Audience development was important for the National Celtic Festival, Stitched-up Textile Festival, and the Bruthen Blues Festival. Strategies to develop audiences that were identified through the mentoring process included broadening the scope of the festivals through the addition of a visual arts component, encouraging the participation of a broad range of local community members in festival programming, offering community-based arts and music activities, and employing market research to identify new audiences.

In the case of the Bruthen Blues Festival, *Directions* accentuated the need to successfully diversify its program, and assisted the mentee to implement strategies aimed at achieving this objective. Bruthen Blues Festival's mentee acknowledged that while there was a need to diversify the program to develop its audiences, she was also cognisant of the need to retain the festival's heritage and established audiences. She said:

"... we have a lot of diehard blues people that come. So we don't want to throw the baby out with the bath water, but we want to expand it [the festival program]... so it's not just the blues. It's still got that element there but there's a whole lot of stuff that's much more community oriented" [Mentee BBAMF].

The National Celtic Festival also recognised opportunities for audience development and, indeed, the Festival's mentee noted that Celtic traditions of art and culture are being interpreted and accepted in Australian communities through a range of media. She believed then that this phenomenon was stimulated by the *Lord of the Rings* which showcased aspects of Celtic culture to Australian communities. Some of these have been embraced by new, non-traditional or fringe markets for Celtic culture and art. This phenomenon provides the National Celtic Festival with opportunities to communicate with new and different audiences. Yet, similar to the thoughts of the Bruthen Blues Festival's mentee, the National Celtic Festival's mentee also recognised that there is a proven market for the festival that holds traditional views of Celtic culture and art. Thus the challenge for the National Celtic Festival is to manage any tensions that emerge between traditional audiences and new audiences.

Similar issues emerged for the Stitched-up Textile Festival. Stitched-up Textile Festival's mentor recognised that difficulties sometimes emerge when balancing festival content with artistic direction. She stated that it would be important for the directors of Stitched-up Textile Festival to 'understand the difference between programming for content as opposed to programming for artistic direction' to ensure the successful artistic direction of the festival. This meant careful diversification, posing some concerns for the mentee. In this instance, diversification not only means broadening the appeal, but also the introduction of exclusive

practices that involve targeting some works, while rejecting others. This was considered to be problematic for a regional festival as previous practice had been to generally accept all the work tendered for inclusion. Participation in *Directions* seemed, however, to assist the mentee to understand the need for such an approach.

Marketing, with an Emphasis on Branding

Branding was particularly important for the Stratford on Avon Shakespearian and the Moss Vale Music Festivals. In the case of the Stratford on Avon Shakespearian, the festival was branded by its location, and while this is useful for providing the festival with a 'sense of place' and Stratford with an identity, it has perhaps contributed to the fact that the festival has not gained strong support from the residents of the local Shire, beyond Stratford. Similarly, the festival's theme places a boundary on it. It was acknowledged that the theme needs to be capitalised upon more effectively in future. Solutions were developed through the mentoring process, which included identifying a broader set of stakeholders for the festival, that is, beyond Stratford and beyond Shakespearians – and actively engaging them in communication.

Both the Moss Vale Music Festival's mentee and mentor recognised the need for the festival to also engage more actively with the local community and its musicians, and although its world music brand was perceived to be a strength, participation in *Directions* assisted the Moss Vale Music Festival's mentee to analyse how the brand was impacting upon the perceptions of the festivals by the local community and local festival musicians. While initially seen as a strength, the 'world music' brand was to some degree stifling the festival's potential to engage with these important stakeholders. Through participation in *Directions*, strategies were developed to capitalise on the brand, so that the festival could more actively engage with local musicians and the community.

Decision-making Skills

As volunteers, many of the mentees lacked confidence to make decisions in an entrepreneurial style. While they were generally confident in making 'safe' decisions, some of the mentees acknowledged that they wanted to develop their decision-making skills to gain a competitive edge for their festivals. Development of decision-making skills was important for National Celtic Festival and Moss Vale Music Festival. Indeed, the mentoring process assisted the National Celtic Festival's mentee to finalise a number of aspects of a three-year strategic plan, particularly with regard to the artistic direction of the festival. She said *Directions* 'really lifted our confidence and enthusiasm and gave us more drive to work on those [artistic] areas' [Mentee NCF]. Similarly, the Moss Vale Music Festival's mentee acknowledged that there was a need to make decisions about the festival more strategically than had been the case in the past. Short-term and long-term decisions were now seen to compliment each other, whereas in the past they were not as well-connected with each other. For example, the Moss Vale Music Festival's mentee suggested said that '... [we] feel more secure that we don't have to get huge amounts of money, that if we are thinking long-term we can do something that aids the artistic development as well as the specific things for that year' [Mentee MVMF].

Financial Considerations

Financial considerations are an important and often over-looked aspect of festival management, including funding and finances.

The Bruthen Blues and Music Festival, for example, has had minimal success gaining funding, but the mentoring process assisted the mentee to identify how the festival's

funding applications could be improved upon in the future. Participation in *Directions* enabled Bruthen Blues and Music Festival's mentee to view the funding application process more strategically, and to recognise the value of developing applications as a team. It also became evident to the mentee that by broadening the artistic direction of the festival, funding sources would also be broadened.

For the National Celtic Festival, a planned organisational restructure, deemed necessary for the sustainability of the festival, would rely on sourcing new funds for its implementation. The mentoring process assisted the festival's mentee to consider alternative financing opportunities to include networking with other festivals so that costs can be defrayed over more than one festival; marketing and merchandising, particularly through the internet and web-based activities; and identifying alternative sources of income, from various government departments and philanthropic organisations.

Human resource management

Human resource management entails recruitment, development, and people motivation for festival success. Festivals for whom this was perceived as particularly important included National Celtic Festival, Apollo Bay Music Festival, and the Stratford on Avon Shakespeare Festival. Both the Apollo Bay Music Festival's mentor and mentee agreed that attaining a balance between the use of paid staff and volunteers is a challenge for the festival. Issues associated with volunteer management centred on the pressures placed upon a small group of individuals running the festival. Although there is a wealth of experience from the business and corporate sectors in this group, gaps remain in its skills-set with regard to artistic direction and event operation management. As a result of *Directions*, the committee has recognised the importance of restructuring the Apollo Bay Music Festival's management. In the short-term, it was decided to create one paid position on the committee. It is not intended, however, that any restructuring of the management should alter the festival's programming or artistic direction. Rather, it was thought that this change would affect the festival at a deeper structural level so that the sustainability of the festival can be enhanced.

Similar issues were raised by the Stratford on Avon Shakespeare Festival's mentee. For this festival the 'problems that were identified were burnout of volunteers, the paid people compared to volunteers and the tensions around that…and … developing a dynamic program and up-skilling the individuals in the management committee who have a real input into program development' [Mentee SASF]. Solutions to these issues included developing strategies to enhance the skills of committee members, or 'up-skilling', increasing levels of skills transfer within and from the committee, and 'outsourcing some of the most complex and time-consuming aspects of the festival' [Mentee SASF]. The committee plans to distinguish those areas of festival management and logistics that can realistically rely on volunteers and those that are more suitable to paid employment to address the '…growing issue of the problem of defining which parts of the festival planning and implementation we attach money to in terms of paying people and which parts are seen as a voluntary component' [Mentee SASF].

Furthermore, formalising the roles and responsibilities of both paid employees and unpaid volunteers is required. These issues relate to both the general management of the festival and to its artistic direction. The committee intends to network and collaborate more actively with other festivals and discuss approaches to human resource management, explore a range of policies, and examine position descriptions. These solutions are associated with the process of succession planning and resource allocation, which are important for any organisation's sustainability. Furthermore, the Stratford on Avon Shakespeare Festival now realises that it competes for funding, both generally and in relation to the arts, against other sophisticated festivals.

As illustrated, not all these areas of management were of concern to all the festivals; however, there was considerable commonality among the festivals with regard to what areas of management are perceived to be of importance to them.

EFFECTIVENESS OF *DIRECTIONS*

This study evaluated the effectiveness of *Directions* and the potential of mentoring within the context of the regional arts festival sector in Victoria, Australia. Importantly, the structure of *Directions* appears to have enabled the mentees and mentors to focus on the issues that were of perceived importance to them, not those that would have been delivered as part of a formal training program. The findings in relation to the effectiveness of *Directions* indicate that mentoring has the potential to be highly useful within the context of regional arts festivals. From this evaluation, mentoring seems to have the potential to assist participants to identify and articulate issues of concern and opportunities for their festivals. Furthermore, and possibly more importantly, mentoring assists participants to develop strategies specific to their needs and objectives that enable them to address their concerns, harness opportunities, and develop skills to operationalise those strategies in the future.

The style of mentoring that developed in the program differed from traditional forms of mentoring in that the mentoring in *Directions* was not always done on a one-to-one basis. Often, the mentors worked with pairs of mentees (from the same festival), and in some cases the mentors worked with entire committees. Furthermore, the goals were not personally oriented; rather they were oriented towards the festivals. These deviations from the traditional notion of mentoring, however, seemed to be appropriate and contributed to the success of the program. The mentors in this program appeared to act as role models, coaches, brokers, and advocates, which Stone (1999) identified as the key attributes of mentoring. They instructed, advised, provided feedback, evaluated and challenged the mentees, and thus can be seen as responding to the recommendations of Zanting *et al.* (2003).

DISCUSSION

This study provides an analysis, through case study examples, of the flexibility of mentoring. *Directions* provided the mentees of the program with professional development that they often find difficult to access in regional locations. Further, it provided the mentees with management frameworks within which to implement their new-found skills. Moreover *Directions* actively transferred skills within the festival sector. As a pilot program, *Directions* was relatively informal and this seemed to suit the nature of the problems encountered in the regional arts festival context. Online communication opportunities may be particularly useful ways to address some of the unique issues that relate to festivals in regional locations. While there is always the temptation to formalise something that is seemingly successful, in this instance, formalising the mentoring program may be counterproductive. *Directions* provided opportunities to develop unique responses to the challenges that each of the festivals were encountering.

This study also highlights that volunteer festival managers perform a diverse range of activities, including strategic planning, human resource management, marketing, funding raising via grant applications, ticketing and merchandising, stakeholder management, festival programming, and tourism development and planning – all while volunteering their services to the festivals, and all set within the context of artistic direction and within a management framework.

Clearly, *Directions* benefited the participants of the program. By disseminating the outcomes of program, other festivals can learn from *Directions*. RAV and AV have already begun to do this at conferences and symposiums for regional festivals. Evaluation of this pilot program provides an opportunity for RAV and AV, in particular, to assess the need for such programs and, as required, develop programs to meet the needs of the burgeoning festival sector in regional Victoria. The underlying principles of *Directions*, where knowledge was disseminated, skills were enhanced, and networks developed in the regional Victorian festival sector in Australia, can be used by a number of other stakeholders in the festival sector, nationally and internationally.

While other research on mentoring has recommended the formalisation of the mentoring programs (cf. Austin *et al.*, 2002), it seems that the key to the success of *Directions* was its informality and its flexibility to respond to the unique needs of the festivals involved. Another reason for the success of *Directions* seems to have been the motivation of those involved to develop relationships with their mentors or mentees. This was particularly obvious at the briefing session because the mentees and mentors demonstrated reciprocal respect for each other and their abilities as artistic directors and festival managers. Hence, Parra *et al.* 's (2002) assertion that 'more positive perceptions of program training and support should predict greater benefits and relationship continuation' has been supported by this study.

FURTHER RESEARCH ON MENTORING AND FESTIVALS

This paper has investigated mentoring within the context of festival management. While mentoring has long been recognised in the arts as a suitable method of training and for knowledge transfer, its application within the context of arts festivals appears to have been limited. In terms of research on mentoring within the context of festivals, this can only be done if researchers are aware of mentoring programs in place within the context of regional arts festivals. It may be that mentoring is being used within this context, but perhaps less formally than was the case for *Directions*. Research to investigate whether this is the case is a worthy line of inquiry. While this study explored the outcomes of *Directions* for the festivals shortly after the completion of the program, the impact of future programs may be more fully evaluated using a longitudinal approach. It may be useful to explore the impact of *Directions* on the festivals, and the participants of the program, in two or three years' time.

In summary, *Directions* was innovative, yet it employed an acceptable and recognisable method of training and transferring knowledge for the arts sector. The method used to evaluate the program was suitable, given the exploratory nature of evaluation itself. It assisted in uncovering the salient themes that are of concern to festival directors in regional areas. Further research on this topic incorporating a larger number of festivals, for example, may find a greater, more diverse, number of these salient themes that are relevant for the festival sector.

ACKNOWLEDGEMENTS

Thanks to Ms Angela Osborne for her research assistance for this paper, and the colleagues at Deakin University who provided feedback on drafts of the paper. Thanks also to Regional Arts Victoria, Arts Victoria, the mentors and mentees who generously gave their time to this project.

REFERENCES

Allen, K. and Shaw, P. (2004) *Festivals Mean Business II: The Shape of Arts Festivals in the UK Update*, London, British Arts Festivals Association.

Alomes, S. (1985) Parades of meaning: the Moomba Festival and contemporary culture. *Journal of Australian Studies*, **17**(November), 3–17.

Anon. (2005) *Year Book Australia 2002: Population-population clock*. Retrieved 17th December, 2005, available at: http://www.abs.gov.au/Ausstats/abs@.nsf/94713ad445ff1425ca256820 00192af2/1647509ef7e25faaca2568a900154b63! OpenDocument

Austin, J., Covalea, L. and Weal, S. (2002) *Going the extra mile – Swinburne University's Mentor Program*, Melbourne.

Booth, M. (1993) The effectiveness and role of the mentor in school: the students view, *Cambridge Journal of Education*, **23**(2), 185–197.

Clark, M. (1981) *The History of Australia,* Vol. 5.

Craik, J., McCallister, L. and Davis, G. (2003) Paradoxes and contradictions in government approaches to contemporary cultural policy: an Australian perspective, *The International Journal of Public Policy*, **9**(1), 17–33.

Davidson, C. and Carlsen, J. (2002) *Event volunteer expectations and satisfaction: a study of protocol assistants at the Sydney 2000 Olympics,* Paper presented at the Annual Council of Australian Tourism and Hospitality Educators Conference, Edith Cowan University.

DuBois, D. L. and Neville, H. (1997) Youth mentoring: Investigation of relationship characteristics and perceived benefits, *Journal of Community Psychology*, **25**(3), 227–234.

Edwards, A. and Collison, J. (1995) What do mentor teachers tell student teachers about pupil learning in infant schools? *Teachers and Teaching: Theory and Pratice*, **1**(2), 265–279.

Elstad, B. (1996) Volunteer perceptions of learning and satisfaction in a mega-event: the case of the XVII Olympic Winter Games in

Lillehammer, *Festival Management and Event Tourism*, **4**(3/4), 75–84.

Elstad, B. (2003) Continuance commitment and reasons to quit: a study of volunteers at a jazz festival, *Event Management: An International Journal*, **8**(2), 99–108.

Formica, S. and Murrmann, S. (1997) The effect of group membership and motivation on attendance: an international festival case, *Tourism Analysis*, **3**(3/4), 197–208.

Freedman, M. (1988) *Partners in Growth: elder Mentors and at Risk Youth*, Philadelphia, Public/Private Ventures.

Frost, W. (2006) From diggers to baristas: tourist shopping villages in the Victorian Goldfields, *Journal of Hospitality and Tourism Management*, **13**(2), 136–143.

Geroy, G. D., Wright, P. C. and Jacoby, L. (2000) Toward a conceptual framework of employee volunteerism: an aid for the human resource manager, *Management Decision*, **38**(4), 280–286.

Hunter, M. A. (2005) Retrieved 14th February, 2005, available at: http://www.ozco.gov.au/arts_resources/publications/getting_connected_making_your_mentorship_work/files/161/getting_connected.pdf+%22Mentoring+in+the+arts%22&hl=en

Jago, L. K. (1998) Special events: a conceptual and differential framework, *Festival Management and Event Tourism*, **5**(1/2), 21–32.

Jago, L. and Deery, M. (2000) The role of human resource practices in achieving quality enhancement and cost reduction: an investigation of volunteer use in tourism organisations, *International Journal of Contemporary Hospitality Management*, **14**(5), 229–236.

Kerstetter, D. L. and Mowrer, P. (1998) Individuals' reasons for attending First Night, a unique cultural event, *Festival Management and Event Tourism*, **5**(3), 139–146.

Lade, C. and Jackson, J. (2004) Key success factors in regional festivals: Some Australian experiences, *Event Management* **9**, 1–11.

Lourdes, B. (1999) The enduring debate over unpaid labour, *International Labour Review*, **138**(3), 287–309.

Parra, G., DuBois, D. L., Neville, H., Pugh-Lilly, A. O. and Povinelli, N. (2002) Mentoring relationships for youth: investigation of a process-oriented model, *Journal of Community Psychology*, **30**(4), 367–388.

Pegg, S. (2002) *Satisfaction of volunteers involved in community events: implications for the event manager.* Paper presented at the Events and Placemaking Conference, Event Research Conference, Sydney.

Ryan, C. (1995) Finance, flowers and festivals – a case study of little economic impact, *Tourism Economics*, **1**(2), 183–194.

Saleh, F. and Wood, C. (1998) Motives of volunteers in multicultural events: the case of the Saskatoon Festival, *Festival Management and Event Tourism*, **5**(1/2), 59–70.

Sipe, C. (2002) Mentoring programs for adolescents: a research summary, *Society for Adolescent Medicine*, **31**, 251–260.

Slaughter, L. (2002) *Motivations of long term volunteers at events.* Paper presented at the Events and Placemaking Conference, Event Research Conference, Sydney.

Sofield, T. H. B. and Li, F., Mei Sarah (1998) Historical methodology and sustainability: an 800 year old festival from China, *Journal of Sustainable Tourism*, **6** (4), 267–292.

Solberg, H. A., Anderson, T. D. and Shibli, S. (2002) An exploration of the direct impacts from business travellers at World Championships, *Event Management: An International Journal*, **7**(3), 151–164.

Spring, J. (1995) *Arts Festivals 1993–94 Research Paper No. 13*, Sydney, Australia Council.

Stone, F. (1999) *Coaching and Counselling and Mentoring*, New York, AMACOM.

Tenner, E. (2004) The pitfalls of academic mentorship, *Chronicle of Higher Education*, **50**, 49.

Uysal, M., Gahan, L. and Martin, B. (1993) An examination of event motivations: a case study, *Festival Management and Event Tourism*, **1** (1), 5–10.

Whitaker, D. M. (2003) The use of full-time volunteers and interns by natural-resource professionals, *Conservation Biology*, **17**(1), 330–333.

Williams, C. C. (2003) Harnessing voluntary work: A fourth sector approach, *Policy Studies*, **23**(3), 247–260.

Williams, M. and Bowdin, G. A. J. (2005) *An investigation into the effectiveness of arts festivals evaluation.* Paper presented at the Impacts of Events: Triple Bottom Line Evaluation and Event Legacies, Sydney.

Williams, P., Dossa, K. B. and Thomkins, L. (1995) Volunteerism and special event management: a case study of Whistler's Men's World Cup of Skiing, *Festival Management and Event Tourism*, **3**(2), 83–96.

Wood, E. (2002) *Events, civic pride and attitude change in a post-industrial town: evaluating the effect of local authority events on residents' attitudes to the Blackburn Region,* Paper presented at the Events and Placemaking Conference. Event Research Conference, Sydney.

Woolf, F. (2004) *Partnerships for Learning: A Guide to Evaluating Arts Education Projects*, London, Arts Council of England.

Young, A. M. and Perrewé, P. L. (2000) What did you expect? An examination of peer-related support and social support among mentors and protégés, *Journal of Management*, **26**(4), 611–632.

Zanting, A., Verloop, N. and Vermunt, J. (2003) Using interviews and concept maps to access mentor teachers practical knowledge, *Higher Education*, **46**, 198–214.

Local food festivals in Northeast Iowa communities: A visitor and economic impact study

Ariana Çela, Jill Knowles-Lankford and Sam Lankford

INTRODUCTION

Tourism, as the largest growing industry, is usually promoted by a country for its ability to spread economic development and reduce inequalities in income distribution. It can generate sales and output, labor earnings and employment, exchange earnings, improvement of balance of payments, important infrastructure development that benefit locals and tourists alike and it can contribute to diversifying the economic base (Durbarry, 2004; Frechtling and Horvath, 1999; Glasson et al., 1995; Lankford and Howard, 1994; Narayan, 2004; Reid, 2003; Oh, 2005). Therefore, efforts to maximise the economic benefits derived from tourism in destination areas have focused on marketing and management strategies to increase the number of tourists, their length of stay and their overall expenditures.

A complimentary way to enhance the benefits of tourism is to expand the economic linkages by increasing the amount of local food used in the industry (Hall, 2005; Telfer and Wall, 1996). The very nature of the food industry lends itself to a marriage with tourism. Food is associated with relaxation, communicating with others, learning about new things, and hospitality (modified from Bruwer, 2003). Food has become an increasingly important element in the tourism industry and up to 25% of total tourist expenditures is accounted for by foods (Hudman, 1986 as cited in Quan and Wang, 2004). Destinations are trying to incorporate local food and beverages in the tourism product, because the consumption of local food and beverages brings the tourists closer to the host culture (Plummer *et al.*, 2005).

Rural communities, in particular, use food festivals to promote local commodities and differentiate themselves from urban community festivals (adapted from Emmons, 2001). Competition for visitors has compelled festival organizers to explore ways and means to increase attendance levels and self-generated revenues. Doing so has required them to look very carefully at their practices and focus on issues such as foods, crafts and customer service. However, very little is known about the characteristics, motivations and satisfaction of visitors to food tourism festivals.

The discretionary nature of expenditures at food festivals necessitates an understanding of visitor's spending behaviours and the underlying factors affecting such behaviour. Understanding expenditure patterns and activities of visitors during their visit to a specific destination is crucial for the strategic planning of events. Community developers and festival professionals can benefit from economic modeling to reveal the impact of tourist spending on the community. This information can help decision makers formulate plans that will help achieve development goals in the community, in particular, for seeking sponsorship for the festivals.

FOOD-BASED TOURISM FESTIVALS

Food and beverage tourism is increasingly being recognized as an important part of the cultural tourism market and particularly a major area of interest for rural regions (Hall and Mitchell, 2001; Hjalager and Richards, 2002). According to Santich (2004), the definition of food tourism includes gastronomic tourism, culinary tourism and cuisine tourism. Wine routes in Italy, Spain, Portugal, France, Australia, New Zealand and California, beer trails in Canada, aboriginal feasts, agritourism activities, cooking schools, farm vacations, farmers' markets, cheese and chocolate factories in Belgium, Switzerland and France are some of the examples of the food and beverage tourism development around the world.

Much research has been conducted on wine tourism (Hall *et al.*, 2000; Hall, 2005; Getz, 2002; Poitras and Getz, 2006; Telfer, 2001) addressing different aspects on demand and supply sides of wine tourism. McBoyle (1996) examined visitor centers at Scottish malt whisky distilleries from the perspective of green tourism. Joliffe (2003) argues, in a discussion about tea attractions, tea tours and tea destinations, that tea tourism can encourage both consumption and the development of relationships with customers. Knowd (2006) explored a different dimension, that of the linkages between tourism and sustainable agriculture, where the primary motivation of the Hawkesbury community in Australia in engaging with harvest tourism was economic sustainability. Telfer and Wall (1996) discussed the complexity of the relationship between food and tourism, where food can be considered as an input as well as an attraction.

Tourists are seeking authentic and unique experiences by consuming local food and beverages (Plummer *et al.*, 2005; Joliffe, 2003). The tourists' interest in local produce may serve to stimulate local awareness and interest and assist not only in diversification and the maintenance of plant and animal variety, but may also encourage community pride and reinforcement of local identity and culture (Hall, 2005; Joliffe, 2003; Quan and Wang, 2004; Sanders, 2005; Szivas, 1999).

Hall and Sharples (2003) presented a framework to understand food tourism as special interest tourism. The travel motivation for this special interest tourism ranges from high interest or gourmet tourism, where the primary motivation is to visit a specific restaurant or winery and the tourism activities are nearly all food related, to low or no interest in food, where tourists visit just a familiar dining establishment because of the need to eat.

Hall (2005), Hall and Sharples (2003), McBoyle (1996) suggested that the relationship between food and tourism represent significant opportunities for rural diversification and regional development. Therefore, many communities see food tourism as an important component in local economic development strategies because of the potential relationships between different industrial sectors, thereby providing not only for a longer circulation of money within local economies, but also the development of new value-added production (OECD, 1995).

Hall (2005) noted several components of such development strategy:

- reduction of economic leakage by using local renewable resources rather than external sources;
- recycling financial resources within the system by buying local goods and services;
- added value to local produce before it is exported;
- connecting local stakeholders to create trust, new linkages and more efficient exchanges;
- attraction of external resources especially finance, skill and technology;
- emphasizing of local identity and authenticity in branding and promotional strategies; and
- selling directly to consumers via farm shops, farmers' market or special events and festivals.

The important component of a food tourism strategy is to showcase the product to tourists. In this respect, special events and festivals, as abovementioned in the strategy components, play an important role. The development of food festivals and other types of festivals has also been fueled by the overall trend towards cultural education

through tourism (Emmons, 2001). From the development standpoint, small festivals in areas with few tourism attractions may be critical in retaining locals' discretionary funds and generating civic pride (Chhabra *et al.*, 2003; Sanders, 2005).

In recent years, special events or festivals have become one of the fastest growing types of tourism attractions (Burr, 1997; Crompton and McKay, 1997; Jago and Shaw, 1998; Thrane, 2002). Tourism marketing professionals increasingly view festivals as integral to tourism development and marketing plans, and they are deliberately creating new festivals as tourist attractions (Getz, 1989; Getz and Frisby, 1988). There has been extensive research focusing on the impacts and outcomes of special events and festivals, some of which are:

- increased visitation to a region (Dwyer *et al.*, 2000; Faulkner *et al.*, 2001; Getz, 1989; Light, 1996; Murphy and Carmichael, 1991; Ritchie, 1984);
- economic injection (Dwyer *et al.*, 2000; Crompton and McKay, 1994; Murphy and Carmichael, 1991; Quinn, 2006; Ritchie, 1984; Ritchie and Smith, 1991; Roche, 1994; Witt, 1988);
- increased employment (Dwyer *et al.*, 2000; Hughes, 1993; Ritchie, 1984);
- improved image of a destination (Backman *et al.*, 1995; Sanders, 2005; Ritchie, 1984; Ritchie and Smith, 1991; Roche, 1994; Witt, 1988);
- enhanced tourism development (Getz, 1989; Hall, 1987; Hughes, 1993; Long and Perdue, 1990; Quinn, 2006; Ritchie, 1984; Ritchie and Smith, 1991);
- ability to act as a catalyst for development (Faulkner *et al.*, 2001; Hughes, 1993; Light, 1996);
- reduction of seasonal fluctuations or extension of the tourism season (Getz, 1989; Ritchie and Beliveau, 1974);
- enhanced community pride (Getz, 1989; Light, 1996; Mules, 1993; Quan and

Wang, 2004; Ritchie, 1984; Roche, 1994; Sanders, 2005); and

- enhanced social and cultural benefits (Dwyer *et al.*, 2000; Gitelson *et al.*, 1995).

In particular, food tourism festivals can also be one alternative opportunity for tourism development in rural areas, adding value to already existing products (Getz and Brown, 2006; Quan and Wang, 2004). Macionis and Cambourne (1998) define necessary pre-requisites for the development of successful food and beverage tourism events as a combination of the active development of food-tourism linkages, a focus on consumer needs and identification of cross-promotional opportunities with the food and tourism sectors. If developed properly, food tourism can add to the range of tourism attractions and provide new attractions to the destination (Szivas, 1999), along with economic benefits.

PURPOSE OF THE STUDY

A wealth of research suggests tangible positive impacts of food and beverage tourism at a community level, if tourism is properly developed and managed. Not only does it enable the development of a 'sense of place', but it also has the potential to generate economic benefits and support sustainable community and tourism development.

Despite many studies investigating the nature and characteristics of special events, continual studies on special community events are needed to demonstrate the economic contribution to the community for more effective decision making (Crompton *et al.*, 2001; McHone and Rungeling, 2000; Wagner, 1997; Wang, 1997). Economic analyses, in particular, supplement the traditional financial balance sheets provided to the government (Crompton *et al.*, 2001), since they address the broader issue of what community residents receive in return for their investment

of tax funds or for their contributions to associations organizing special events.

Also, the economic impact studies help sponsors determine their level of support based on the event's financial merit. Because visitor spending signifies the event's visibility and popularity, funding or corporate sponsorship can be sought with convincing rationale (Irwin *et al.*, 1996). Visitor data obtained from economic impact analyses provide also valuable market information to be further utilized for other management purposes. For example, attendance count, demographic characteristics and spending patterns can serve as the basis for future event planning, target market segmentation and sponsorship solicitation (Yoon *et al.*, 2000; Irwin *et al.*, 1996).

This study, in particular, analyzes travel motivations of visitors, their common interests in local foods in Northeast Iowa community festivals and their economic impact in the community. Food is being considered local if it is being grown, or processed locally. A comparison of the results with visitor data from the same region was undertaken in order to identify the unique characteristics of local food tourism. Specially, the study identifies and describes:

- festival participants' profile;
- festival participants' travel motivations;
- sources of travel information; and
- the economic impact of local food tourism on the communities.

STUDY SITE AND METHODOLOGY

Visitors were intercepted at 11 festivals taking place in Northeast Iowa with a theme of a local food/product of the surrounding region from May–October 2005. Table 1 shows the list of the festivals with the dates and locations.

The list of festivals was prepared based on a web search of Northeast Iowa counties websites and it was screened based on whether

Table 1 List of Festivals

No.	Festivals	Dates of festivals	Place	County	No. of respondents
1.	Dairy parade	June 6th	Waukon	Allamakee	28
2.	Dairy days	June 7th, 8th	Fredericksburg	Chickasaw	27
3.	Strawberry days	June 10th, 11th, 12th	Strawberry Point	Clayton	25
4.	Corn days	August 5th, 6th	Dows	Wright	6
5.	Watermelon days	August 6th	Atkins	Benton	7
6.	Sauerkraut days	August 11th, 12th, 13th	Lisbon	Linn	11
7.	Sweet corn days	August 13th	Elkader	Clayton	13
8.	Watermelon days	August 27th	Fayette	Fayette	9
9.	Sauerkraut days	Sept 2nd, 3rd, 4th	Blairstown	Benton	19
10.	Annual honey fest	September 25th	Indian Creek Nature Center	Linn	19
11.	Apple festival	October 9th	Le Claire	Scott	16

the festivals highlighted a local food theme. Different festivals total estimated attendees ranged from 100 to 1000 visitors during the entire scheduled timeframe. By October 2005, 180 surveys were completed. The respondents have been randomly selected (every one of three attendees) at each of 11 festivals site. Table 1 also shows data on the number of respondents at each festival site.

A questionnaire was designed for the purpose of this study. In the first part of the questionnaire, visitors were asked about travel and trip characteristics, which included questions on the number of previous visits to the region and other festivals they had visited, the primary purpose of the trip, where they had gathered information about the festival, length of stay, type of accommodation and travel party size. In the second part of the questionnaire, they were asked about their spending patterns for lodging, restaurants, groceries, transportation, admissions, shopping and particularly purchasing local food at the festivals, as well as their willingness to pay more for local food. In the third part visitors were asked about their motivations for

attending local food festivals and also about their satisfaction with the food and their overall visit. In the fourth part of the questionnaire, questions on visitors' demographic profile were included, such as age, income, gender, education level, and residence.

RESULTS

The trip characteristics, motivation and satisfaction, demographics, spending patterns and economic impact of the visitors in local food festivals were compared, where it was possible, with data for the visitors from the same area, provided from a study of visitor characteristics of Silos and Smokestacks National Heritage Area (Sustainable Tourism and Environment Program, 2004), in Northeast Iowa.

Trip characteristics

Most of the visitors to the local food festivals were repeat visitors, the same as visitors to Silos and Smokestacks National Heritage Area (SSNHA) and they were visiting specifically to attend the festival, which they learned about primarily by word of mouth. The majority of repeat visitors (54%) have visited the region between two and five times and 22% of them have visited the region six to ten times. Interestingly, a higher percent of festival visitors were repeat visitors.

The visitors to local food festivals stay overnight with their friends or relatives in the area, while visitors to SSNHA stay in a hotel or motel. They are predominantly on a day trip, while SSNHA visitors are more often on an overnight trip. Amongst those visitors in local food festivals who stayed overnight, stayed on an average 1.4 nights, while SSNHA visitors stayed 4.8 nights. Visitors to local food festivals, like visitors to SSNHA, are not in an organized group (Table 2).

Motivation and Satisfaction

Visitors responded to a list of motivations to attend the festival based on a five-point Likert scale, varying from 5 = Strongly Agree to 1 = Strongly Disagree. Factor analysis (Principal Components Analysis with Varimax Rotation) was utilized to better understand the motivation dimensions (Table 3). Three components account for 66.73% of the variability of the original list of motivations.

Factor 1 consists of 'to taste local/fresh foods, to taste food easily available in my hometown, to purchase organic food, to purchase local foods, to support local producers and to connect to a sense of community and place'. This factor is named as motivations to support, taste and purchase local food, and

Table 2 Trip Characteristics of Visitors in Local Food Festivals Compared with SSNHA

	Trip characteristics	Local food festivals %	SSNHA %
Previous visits	First time visitors	15	36.3
	Repeat visitors	85	63.7
Primary purpose of the trip	Specifically to attend the festival/festival	63	16.1
	Friends/relatives	28	22.9
	Side trip	3	19.3
	Business	6	8.8
	Specifically to visit the heritage site	n.a.	31.2
Sources of information	Word of mouth	56	n.a.
	Newspaper	28	14.3
	Other	9	12.3
	Website	3	12.5
	Brochure	1	4.5
	Magazine	1	5.7
	Convention bureau	1	20.6
	Travel publication	1	11.2
Day/overnight trip	Day trip	88	49.2
	Overnight trip	12	50.8
Place of accommodation	Hotel or motel	7.8	22.4
	Bed and breakfast	1.4	1.9
	Staying with friends and relatives	13.3	18.5
	Campground	4.1	9.3
	Staying outside the area	2.7	4.4
Organized group	Yes	2	8
	No	98	92

Table 3 Rotated Component Matrix of Motivations to Attend the Local Food Festival

Motivations	Component		
	1	2	3
To purchase local foods	0.840		
To taste food easily available by my hometown	0.742		
To purchase organic food	0.715		
To taste local/fresh foods	0.626		
To support local producers	0.566		
To connect to a sense of community and place	0.556		
To relax		0.856	
To enjoy the scenery		0.809	
To have a good time with friends and family		0.790	
To learn about food traditions of the region			0.861
To learn about the food-producing process			0.789
To learn about new things in Northeast Iowa			0.689

Extraction method: Principal component analysis.
Rotation method: Varimax with Kaiser normalization.
A rotation converged in five iterations.

accounts for 25.81% of the variance amongst all the lists of motivations.

Factor 2 consists of 'to relax, to enjoy the scenery and to have a good time with friends and family', which is named as motivations to attend the festival, and accounts for 21.77% of the variance.

Factor 3 consists of 'to learn about the food traditions of the region, to learn about the food-producing process and to learn new things in Northeast Iowa' which is named as motivations to learn about local food and Northeast Iowa and accounts for 19.16% of the variance.

Most visitors were motivated to simply attend the festival ($M = 4.59$), followed by the motivation to support, taste and purchase local food ($M = 4.37$) and to learn about local food ($M = 3.97$) (Table 4). They were also satisfied with the visit ($M = 4.40$) and with the local food (4.31).

Demographics

Visitors to local food festivals were in the 26–35 years old category (on an average 40

years old) and they were predominantly from Northeast Iowa (Table 5). SSNHA visitors are older, on an average 51 years old. Visitors were middle income and college educated, similar to SSNHA visitors.

Spending Patterns

The highest spending category for visitors attending the festivals is lodging ($18.46 per person a day), followed by shopping ($8.47 per person a day) and restaurants ($8.32 per person a day) (Table 6). Average party size was 2.65, with 1.4 average overnights. Spending per travel party for festivals attendees was $70.04 or $56.66 per person as compared to SSNHA visitors $ 236.30 per party size (mean party size = 2.85 with an average of 4.8 nights) or $33.63 per person. Interestingly, per person spending for festival attendees, for total and all spending categories, was significantly higher than per person spending for SSNHA visitors.

Visitors to local food festival spent $4.54 per person a day on purchasing locally produced foods.

Table 4 Motivations and Satisfaction with Local Food Festivals

Motivations and satisfaction		Level of agreement/ satisfaction	Percentage	Mean score
Motivations to support, taste and purchase local food (M = 4.37)	To purchase local foods	Agree	50.9	4.04
	To taste food easily available by my hometown	Agree	50.9	4.00
	To purchase organic food	Agree	43.7	3.72
	To taste local/fresh foods	Agree	53.7	4.25
	To support local producers	Agree	49.7	4.24
	To connect to a sense of community and place	Agree	43.8	4.08
Motivations for attending the festival (M = 4.59)	To relax	Strongly agree	54.6	4.48
	To have a good time with friends and family	Strongly agree	53.0	4.48
	To enjoy the scenery	Strongly agree	46.6	4.40
Motivations to support and learn about local food (M = 3.97)	To learn about the food traditions of the region	Agree	49.04	3.76
	To learn about new things in Northeast Iowa	Agree	4.3	3.81
	To learn about the food-producing process	Agree	42.4	3.65
Overall satisfaction	Overall satisfaction with the visit	Satisfied	47.4	4.40
	Overall satisfaction with the local food	Satisfied	52.3	4.31

5 = Strongly agree, 4 = agree, 3 = neutral, 2 = disagree, 1 = strongly disagree.

Although not included in Table 6, a *t*-test (independent samples) was conducted to find out if there was a difference in spending means for first time and repeat visitors, and between visitors those have as their primary purpose of the trip specifically to attend this festival and those to visit their friends and relatives. First time visitors (M = 96.40, SD = 51.233) did not spend significantly more on lodging than repeat visitors (M = 48.57, SD = 30.648), at p = 0.069, t(10) = 2.034. Statistically insignificant results were also attained in differences in other spending categories between first time and repeat visitors and between visitors travelling specifically to attend the festival and those to visit friends and relatives.

Differences in mean spending for groceries between visitors with household incomes less than \$34,999 and visitors with household incomes \$35,000–\$75,999 were statistically significant at p = 0.027, F(2,19) = 4.392. Differences in mean spending of other categories were not statistically significant.

Thirty-six percent of the visitors were willing to pay 1–5% more for the local food, the same as there were to pay 6–10% more and 15% of the visitors are willing to pay more than 20% more for the local food. Only 8% are not willing to pay more for this type of product.

Table 7 presents the cross tabulation between willingness to pay, and household income, age group and gender of the visitors. Spearman *rho*, ρ was used to determine the significance of differences amongst above mentioned categorical and ordinal variables.

Table 5 Visitors Demographics in Local Food Festivals as Compared with SSNHA Visitors

Visitors' demographics		Local food %	SSNHA %
Residency	Iowa	98.8	n.a.
	Illinois	1.2	n.a.
Age	25 and under	13.9	5.8
	26–35 years	31.1	12.2
	36–45 years	24.5	20.0
	46–55 years	12.6	21.5
	56–65 years	9.3	23.1
	66–75 years	4.6	11.8
	76–95 years	4.0	5.5
Gender	Female	52.3	56.5
	Male	47.7	43.5
Income level	less 9,999	1.9	3.9
	10,000–14,999	2.9	1.3
	15,000–24,999	6.7	5.4
	25,000–34,999	16.3	10.3
	35,000–49,999	16.3	16.7
	50,000–74,999	22.1	22.7
	75,000–99,999	13.5	11.1
	100,000–149,000	7.7	9.4
	150,000–199,999	0	1.9
	200,000 and above	1.9	1.5
	Chose not to answer	10.6	15.8
Education level	Some High School	5.3	1.4
	High School Graduate	22.1	16.1
	Some College	26.0	25.9
	College Graduate	33.6	26.3
	Post Graduate Work	5.3	7.2
	Post Graduate Degree	6.9	18.8
	Technical School	0.8	2.3

Most of the visitors with household income below \$34,999 were willing to pay 1–5% more for local food, while visitors with income above \$75,000 were willing to pay 6–10% more. However, willingness to pay is not associated with the household level of income (Spearman $\rho = 0.023$, $p = 0.834$). Visitors below 45 years old were willing to pay 1–5% more on the price for the local food, while visitors 46–65 years old were willing to pay more than 20% on the price of the local food. The willingness to pay more for local food is significantly correlated with visitors age (Spearman $\rho = 0.211$, $p = .011$). Most of the women are willing to pay 6–10% more for local food, while men are more willing to pay 1–5% more. Willingness to pay more for local food and gender of the visitors are significantly correlated (Spearman $\rho = -0.224$, $p = 0.007$).

Economic Impact of Local Food Festivals

The IMPLAN model utilized for the purpose of this study estimates the economic impact of local food festivals in Allamakee, Benton, Chickasaw, Clayton, Fayette, Linn,

Table 6 Mean of Spending Categories of Visitors in Local Food Festivals Compared with SSNHA

Spending categories	Local Food Festivals		SSNHA	
	$ (Mean party size = 2.65; Mean nights = 1.4)	$ Per person a day	$ (Mean party size = 2.85; Mean nights = 4.8)	$ Per person a day
Lodging	68.50	18.46	175.4	12.82
Restaurant	30.86	8.32	73.5	5.37
Groceries	19.64	5.29	41.6	3.04
Transportation expense	26.38	7.11	45.5	3.33
Admissions	16.52	4.45	37.7	2.76
Shopping	31.42	8.47	86.4	6.32
Purchasing locally produced foods	16.86	4.54	n.a.	n.a.
Total spending	70.04	56.66	236.3	33.63

Scott, Wright and their contiguous counties (30) in Northeast Iowa. The area is 22,329 square miles, has a population of 1,327,206 and 565,795 households (IMPLAN model).

IMPLAN input–output model was originally developed by the US Forest Service, and currently maintained by the Minnesota IMPLAN Group. Input–output models estimate the increase in economic activity associated with some money injection such as visitor expenditure in the economy. It shows the uses of the output from each sector or industry as an input to other industries/

Table 7 Cross Tabulation of Willingness to Pay More with Visitors' Household Income, Age and Gender (Percentage)

		Willingness to pay more						Spearman ρ
		no more	1–5%	6–10%	11–20%	more than 20%	Total	
Income	Below $34,999	14.8	33.3	29.6		7.4	14.8	100	0.023
	$35,000–$74,999	10.8	32.4	32.4	8.1	16.2	100	
	Above $75,000	4.2	33.3	54.2	8.3	0	100	
	Total	10.2	33.0	37.5	8.0	11.4	100	
Age Group	Below 25 years	4.8	61.9	23.8	4.8	4.8	100	0.211*
	26–35 years	8.9	46.7	31.1	4.4	8.9	100	
	36–45 years	8.3	33.3	27.8	11.1	19.4	100	
	46–55 years	0.0	16.7	33.3	5.6	44.4	100	
	56–65 years	15.4	7.7	38.5	0.0	38.5	100	
	66–75 years	14.3	57.1	28.6	0.0	0.0	100	
	Above 76 years	25.0	25.0	50.0	0.0	0.0	100	
	Total	8.3	38.2	30.6	5.6	17.4	100	
Gender	Female	6.7	28.0	33.3	9.3	22.7	100	−0.224*
	Male	10.1	44.9	30.4	2.9	11.6	100	
	Total	8.3	36.1	31.9	6.3	17.4	100	

*Significant at $p < 0.05$

sectors of the economy. It is basically a matrix where sectors in one axis represent the suppliers to the sectors on the other axis, which, from the other side, represent the demanders. Input–output analysis today is the most frequently used approach to evaluate the economic impacts of tourism, primarily on consumption, income and employment (Crompton *et al.*, 2001; Fletcher, 1989; Frechtling and Horvath, 1999).

The estimation of the total economic impact in Northeast Iowa of eleven local food festivals involves the production of three different estimates.

- The estimated effect of visitor spending on the total value of economic transactions in the economy of Allamakee, Benton, Chickasaw, Clayton, Fayette, Linn, Scott Wright and their 30 contiguous counties in Iowa.
- The estimated effect of visitor spending on the overall level of household income in the above mentioned counties.
- The estimated effect of visitor spending on the number of jobs in the above-mentioned counties.

Table 8 presents the economic effects associated with visitors spending at the festivals. The total estimated number of visitors in the local food festivals, based on the assessment of festival organizers, was 36,200. Due to the lack of information of local food festivals substitutability by the visitors, for the purpose of assessing the economic impact of local food festivals, only the number of visitors whose primary purpose of visit was specifically to attend the festival was taken into account. Therefore, the initial visitor spending was estimated from visitor surveys based on an assessed number of visitors of 22,806 (36,200 × 63%).

Initial visitor spending is just over $1.6 million and are identified as total 'Direct' economic transactions in Table 8. Direct

spending is the input or the injections in the economy of the above-mentioned Northeast Iowa counties that will be multiplied further, based on the linkages of different economic sectors in the area.

Table 8 presents the indirect and induced effects, in terms of the value of the economic transactions that result from the multiplier process. Indirect effects measure the total value of supplies and services supplied to tourism-related businesses by the chain of businesses which serve tourism-related businesses. Induced effects accrue when tourism-related businesses and businesses in the indirect industries spend their earnings (wages, salaries, profits, rent and dividends) in goods and services in the area. Total economic impacts are the sum of direct, indirect and induced effects, and are the total of transactions attributable directly to the local food festivals.

Overall, an estimated $2.6 million of gross sales transactions are directly or indirectly related to local food festivals in Northeast Iowa counties, implying an output or gross sales multiplier of 1.61 (2,638,811/ 1,642,394). So, an additional tourist spend of $1 generates $1.61 of output in the economy of selected Northeast Iowa counties. Nearly $465,000 of these effects is 'indirect', thus representing the supply transactions that support tourism-related businesses. Approximately $530,000 of these effects is 'induced', thus resulting from personal purchases made by workers of tourism-related businesses and indirect businesses in the area.

Table 8 also presents that the personal income component of 'total' spending is above $1.4 million and the estimated number of jobs created in selected Northeast Iowa counties from local food tourism festivals taking place is 51 jobs. The income multiplier is 1.65 (1,413,309/857,234) and the employer multiplier is 1.29 (51.1/39.6). Since serving and organizing local food festivals is primarily seasonal and a secondary or

Table 8 The Annual Economic Impact of Visitors to Local Food Festivals 2005 ($Sales, $Value Aded, Jobs)

Industry	Sales/output				Total value-added/income	Total jobs/employment
	Direct	Indirect	Induced	Total		
Agriculture	0	11,258	7,579	18,837	3,144	0.1
Mining	0	9	11	21	11	0.0
Utilities	0	23,633	13,286	36,920	21,550	0.1
Construction	0	14,042	3,579	17,621	7,085	0.2
Manufacturing	0	89,323	75,807	165,134	64,249	0.6
Transportation and warehousing	0	29,713	13,019	42,733	23,651	0.5
Retail trade	752,112	21,230	70,311	843,651	411,239	19.5
Information services	0	42,917	14,877	57,797	29,798	0.4
Finance, insurance and real estate	1,940	84,418	69,620	155,980	103,052	1.4
Professional services	3	77,990	21,789	99,778	70,962	1.8
Other services	859,841	58,225	170,336	1,088,400	617,013	26.4
Government	28,498	12,174	71,266	111,939	61,555	0.1
Total	1,642,394	464,932	531,480	2,638,811	1,413,309	51.1

Source: IMPLAN model for eight Northeast and their contiguous counties in Iowa.

volunteer occupation, the estimate of 'direct' jobs associated with local food festivals (approximately 51 jobs) should be cautiously interpreted. Direct jobs refer to the jobs that are generated directly from the measured activity. Therefore, the direct involvement in organizing and providing local food festivals cannot be deemed as employment directly generated by the local food festivals.

The services groupsector was generating $1 million in industrial output; $617,013 in value added impact and supported 26 jobs. This is 51.7% of the total employment attributed to local food festivals in Northeast Iowa. Importantly trade aggregate contributed in output $844,000; in value added benefits $411,000 and in employment with 20 jobs.

Table 9 shows a comparison of the spending impact of visitors to local food festivals and in the SSNHA. The impacts of spending from SSNHA visitors are substantially higher, since the assessment is based on an estimated number of 547,773 visitors to SSNHA, much higher than 22,806 visitors to local food festivals. However, the multipliers are showing similar impact of local food festivals with heritage sites of SSNHA.

SUMMARY AND CONCLUSIONS

Visitors to local food festivals in Northeast Iowa were typically middle aged, relatively younger than SSNHA heritage visitors. They were college graduates and affluent, similar to SSNHA visitors. Festivals visitors, like SSNHA heritage visitors, were predominantly repeat visitors and not part of an organized group. Festivals attendees were mainly day trippers, unlike SSNHA heritage visitors. They were primarily motivated to specifically attend the festivals, closely followed by the motivation to support, taste and purchase local food. So they may be considered as high interest or gourmet tourists according to Hall and Sharples (2003) spectrum of

Table 9 The Annual Economic Impact of Visitors to Local Food Festivals Compared with SSNHA Visitors ($Sales, $Value Added, Jobs)

Impact	Direct impact		Total impact		Multipliers	
	Local food festivals visitors	SSNHA visitors	Local food festivals visitors	SSNHA visitors	Local food festivals visitors	SSNHA visitors
Sales/output impact ($)	1,642,394	62,164,637	2,638,811	102,928,867	1.61	1.66
Value added/income Impact($)	857,234	33,361,234	1,413,309	56,478,159	1.65	1.69
Jobs Impact	39.6	1,491.5	51.1	1,981.4	1.29	1.33

food tourism motivations. Visitors to local food festivals had an impact in terms of output or sales of $2.6 million, and for every dollar spent $0.61 in new output was created (output multiplier = 1.61), which is significant and rather comparable with the impacts of SSNHA heritage sites in the Northeast Iowa communities.

Findings support the existing literature focusing on the impacts and outcomes of special events. Local food festivals in Northeast Iowa contributed economically in the area by generating more output, income and jobs. The economic impacts of the festivals were widely distributed across all sectors of the economy, thus they were acting as a catalyst for development in Northeast Iowa. Interestingly, the economic impacts of local food festivals, in terms of the multiplier, were similar to heritage sites. Thus, the development of local food tourism in Northeast Iowa constitutes an important component in the regional economic development strategy.

From the analysis of visitor characteristics, local food festivals attendees constitute a similar but distinct market from heritage visitors. Local food festivals attendees are younger, spend more per person per day and stay for shorter periods in the area than heritage visitors.

Findings also revealed some practical implications for festival organizers. Advertisements should be put in place to attract more first time visitors and younger and older generations. Partnerships and collaboration with other tourism and tourism-related businesses is critical to enhance larger and longer visitation in the region. In this perspective, networks and cluster relationships, as suggested by Hall (2005), are a significant part of the development of 'intangible capital' for host communities, enabling them to excel in the exercise of food tourism activities. In this regard, the issue of networks and cluster relationships could constitute a future research topic to be addressed and thoroughly elaborated.

In addition, exploring the link between food tourism and sustainable agriculture can be deemed a future research topic. Attaining a secure future for agriculture in Northeast Iowa also has potential for regional identity and tourism. The key element in agriculture and tourism's interest is in maintaining the landscape. By ensuring viable agriculture, communities can ensure its contribution to the rural character of place and its role in conserving and protecting the commons. It also creates new potential in product development for food-based tourism.

REFERENCES

Backman, K., Backman, S., Uysal, M. and Mohr Sunshine, K. (1995) Festival tourism: an examination of motivations and activities, *Festival Management & Event Tourism*, **3**(1), 15–24.

Bruwer, J. (2003) South African routes: some perspectives on the wine tourism industry's structural dimensions and wine tourism product, *Tourism Management*, **24**(4), 423–435.

Burr, S.T. (1997) Love thy neighbor and prosper: community festivals and festivals, *Parks & Recreation*, **32**(3), 106–118.

Chhabra, D., Sills, E. and Cubbage, F. W. (2003) The significance of festivals to rural economies: estimating the economic impacts of Scottish Highland Games in North Carolina, *Journal of Travel Research*, **41**(40), 421–427.

Crompton, J. L. and McKay, S. (1994) Measuring the economic impact of festivals and festivals: some myths, misapplications and ethical dilemmas, *Festival Management and Event Tourism*, **2**(1), 33–43.

Crompton, J. L. and McKay, S. (1997) Motives of visitors attending festival festivals, *Annals of Tourism Research*, **24**(2), 425–439.

Crompton, J. L., Lee, S. and Shuster, T. J. (2001) A guide for undertaking economic studies: the Springfest example, *Journal of Travel Research*, **40**(1), 79–87.

Durbarry, R. (2004) Tourism and economic growth: the case of Mauritius, *Tourism Economics*, **10**, 389–401.

Dywer, L., Mellor, R., Mistilis, N. and Mules, T. (2000) A Framework for assessing "tangible" and "intangible" impacts of festivals and conventions, *Event Management*, **6**(3), 175–189.

Emmons, N. (2001) Festivals are salivating while celebrating food, *Amusement Business*, **113**(1), 24–26.

Faulkner, B., Chalip, L., Brown, G., Jago, L., March, R. and Woodside, A. (2001) Monitoring the tourism impacts of the Sydney 2000 Olympics, *Event Management*, **6**(4), 231–246.

Fletcher, J. (1989) Input-output models, in: T. Baum and R. Mudambi (eds.) *Economic and management methods for tourism and hospitality research*. Chichester, UK, John Wiley & Sons.

Frechtling, D. C. and Horvath, E. (1999) Estimating the multiplier effects of tourism expenditures on a local economy through a regional input-output model, *Journal of Travel Research*, **37**(4), 324–332.

Getz, D. (1989) Special festivals: Defining the product, *Tourism Management*, **10**(2), 125–137.

Getz, D. (2002) *Explore Wine Tourism: Management, Development and Destinations*, New York, Cognizant Communication Corporation.

Getz, D. and Brown, G. (2006) Critical success factors for wine tourism regions: a demand analysis, *Tourism Management*, **27**(1), 146–158.

Getz, D. and Frisby, W. (1988) Evaluating management effectiveness in community-run festivals, *Journal of Travel Research*, (Summer), 22–27.

Gitelson, R., Kerstetter, D. and Kiernan, N. E. (1995) Evaluating the educational objectives of a short-term festival, *Festival Management & Event Tourism*, **3**(1), 9–14.

Glasson, J., Godfrey, K. and Goodey, B. (1995) *Towards visitor impact management: Visitor impacts, carrying capacity and management responses in Europe's historic towns and cities*. England, Avebury.

Hall, C. (1987) The Effects of hallmarks festivals on cities, *Journal of Travel Research*, **26**(2), 44–45.

Hall, C. M. (2005) Rural wine and food tourism cluster and network development, in: D. Hall, I. Kirkpatrick and M. Mitchell (eds.) *Rural Tourism and Sustainable Business*, Clevedon, England, Channel View Publications, pp. 149–164.

Hall, C. M. and Mitchell, R. (2001) Wine and food tourism, in: N. Douglas and R. Derret (eds.) *Special Interest Tourism: Context and Cases*, Brisbane, Australia, John Wiley & Sons, pp. 307–329.

Hall, C. M. and Sharples, L. (2003) The consumption of experiences or the experience of consumption? An introduction to the tourism of taste, in: C. M. Hall, L. Sharples, R. Mitchell, N. Macionis and B. Cambourne (eds.) *Food Tourism Around the World: Development, Management and Markets*, Oxford, Butterworth Heinemann, pp. 1–24.

Hall, C. M., Sharples, L., Cambourne, B., Macionis, N., Mitchell, R. and Johnson, G. (eds.) (2000) *Wine Tourism around the World: Development, Management and Markets*, Oxford, Butterworth Heinemann.

Hjalager, A. and Richards, G. (eds.) (2002) *Tourism and Gastronomy*, London, Routlege.

Hughes, H. (1993) Olympic tourism and urban regeneration, *Festival Management & Event Tourism*, **1**(4), 157–162.

Irwin, R. L., Wang, P. and Sutton, W. A. (1996) Comparative analysis of diaries and projected spending to assess patron expenditure behavior at short-term sporting events, *Festival Management & Event Tourism*, **4**(1/2), 29–37.

Jago, L. K. and Shaw, R. N. (1998) Special festivals: a conceptual and definitional framework, *Festival Management & Event Tourism*, **5**(1), 21–32.

Joliffe, L. (2003) The lure of tea: history, traditions and attractions. In C. M. Hall, L. Sharples, R. Mitchell, N. Macionis and B. Cambourne (eds.) *Food Tourism Around the World: Development, Management and Markets*, Oxford, Butterworth Heinemann, pp. 121–136.

Knowd, I. (2006) Tourism as a mechanism for farm survival, *Journal of Sustainable Tourism*, **14**(1), 24–42.

Lankford, S. V. and Howard, D. R. (1994) Developing a tourism impacts attitudes scale, *Annals of Tourism Research*, **21**(1), 121–139.

Light, D. (1996) Characteristics of the audience for "festivals" at a heritage site, *Tourism Management*, **17**(3), 183–190.

Long, P. and Perdue, R. (1990) The Economic impact of rural festivals and special festivals: Assessing the spatial distribution of expenditures, *Journal of Travel Research*, **28**(4), 10–14.

Macionis, N. and Cambourne, B. (1998) Wine and food tourism in the Australian capital territory: exploring the links, *International Journal of Wine Marketing*, **10**(3), 5–23.

McBoyle, G. (1996) Green tourism and Scottish distilleries, *Tourism Management*, **17**(4), 255–263.

McHone, W. and Rungeling, B. (2000) Practial issues in measuring the impact of a cultural tourist event in a major tourist destination, *Journal of Travel Research*, **38**(3), 299–302.

Mules, T. (1993) A Special festival as part of an urban renewal strategy, *Festival Management and Event Tourism*, **1**(2), 65–67.

Murphy, P. E. and Carmichael, B. A. (1991) Assessing the tourism benefits of an open access tournament: the 1989 B.C. winter games, *Journal of Travel Research*, **29**(3), 32–36.

Narayan, P. K. (2004) Economic impact of tourism on Fiji's economy: empirical evidence from computable general equilibrium model, *Tourism Economics*, **10**(4), 419–433.

OECD (Organisation for Economic Cooperation and Development) (1995). *Niche Markets as a Rural Development Strategy*, Paris, OECD.

Oh, C. O. (2005) The contribution of tourism development to economic growth in the Korean economy, *Tourism Management*, **26**(1), 39–44.

Plummer, R., Telfer, D., Hashimoto, A. and Summers, R. (2005) Beer tourism in Canada along the Waterloo-Wellington Ale trail, *Tourism Management*, **26**(3), 447–458.

Poitras, L. and Getz, D. (2006) Sustainable wine tourism: The host community perspective, *Journal of Sustainable Tourism*, **14**(5), 425–448.

Quan, S. and Wang, N. (2004) Towards a structural model of the tourist experience: an illustration from food experiences in tourism, *Tourism management*, **25**(3), 297–305.

Quinn, B. (2006) Problematising "festival tourism": arts festivals and sustainable development in Ireland, *Journal of Sustainable Tourism*, **14**(3), 288–306.

Reid, D. G. (2003) *Tourism, globalization and development: responsible tourism planning*, London, Pluto Press.

Ritchie, J. (1984) Assessing the impact of hallmark festivals: conceptual and research issues, *Journal of Travel Research*, **23**(1), 2–11.

Ritchie, J. and Beliveau, D. (1974) Hallmark festivals: an evaluation of a strategic response to seasonality in the travel market, *Journal of Travel Research*, **13**(2), 14–20.

Ritchie, J. and Smith, B. (1991) The impact of a mega-festival on host region awareness: a longitudinal study, *Journal of Travel Research*, **30**(1), 3–10.

Roche, M. (1994) Mega-festivals and urban policy, *Annals of Tourism Research*, **21**(1), 1–19.

Sanders, K. J. (2005) *Factors of success in northeast Iowa community festivals*. Unpublished master's thesis, University of Northern Iowa, Cedar Falls, USA.

Santich, B. (2004) The study of gastronomy and its relevance to hospitality education and training, *International Journal of Hospitality Management*, **23**(1), 15–24.

Sustainable Tourism and Environment Program (2004) Silos and Smokestacks National Heritage Area Economic Impact and Visitor Study. Retrieved, November 18th, 2005, available at: http://www.uni.edu/step/projects.htm

Szivas, E. (1999) The development of wine tourism in Hungary, *International Journal of Wine Marketing*, **11**(2), 7–18.

Telfer, D. J. (2001) Strategic alliances along the Niagara Wine Route, *Tourism Management*, **22**(1), 21–30.

Telfer, D. J. and Wall, G. (1996) Linkages between tourism and food production, *Annals of Tourism Research*, **23**(3), 635–653.

Thrane, C. (2002) Jazz festival visitors and their expenditures: linking spending patterns to musical interest, *Journal of Travel Research*, **40**(3), 281–286.

Wagner, J. E. (1997) Estimating the economic impacts of tourism, *Annals of Tourism Research*, **24**(3), 592–608.

Wang, P. (1997) Economic impact assessment of recreation services and the use of multipliers: a comparative examination, *Journal of Park and Recreation Administration*, **15**(2), 32–43.

Witt, S. (1988) Mega-events and mega-attractions, *Tourism Management*, **9**(1), 76–77.

Yoon, S., Spencer, D. M., Holecek, D. F. and Kim, D. K. (2000) A profile of Michigan's festival and special event tourism market, *Event Management*, **6**(1), 33–44.

Festival evaluation: An exploration of seven UK arts festivals

Michael Williams and Glenn A J Bowdin

INTRODUCTION

Within the United Kingdom (UK) arts sector, which includes festivals, stakeholders are increasing demands for accountability in order to justify funding, while those involved in delivery are developing appropriate responses. Alongside this, interest in arts festivals research is increasing as the events subject area develops and matures, yet, to date, there appears to be a paucity of previous research into arts festivals in general, and arts festivals evaluation in particular.

This study examines the significance of evaluation in terms of arts festival management and the process of evaluation. The primary aim is to establish the approaches taken for evaluation in arts festivals. The paper commences with a brief overview of arts festivals in the UK and explores evaluation literature and research to place the study in context. It then moves on to present the findings of a study into the approaches of seven arts festivals in the United Kingdom.

CONTEXT

The term 'festival' has been used for hundreds of years and has been used to cover a multitude of events (Bowdin, Allen, O'Toole, McDonnell and Harris, 2006). Getz (2005, p. 21) provides a succinct working definition of a festival, defining it as 'a public, themed celebration', while Smith (1990, p. 128) provides a more detailed definition, perhaps more useful for the purposes of this study, identifying festival as, 'a celebration of a theme or special event for a limited period of time, held annually or less frequently (including one-time only events), to which the public is invited.'

Due to the diversity in types of art forms and festivals, and the comparatively limited research into arts festivals, the specific definition of an 'arts festival' is also open to discussion. However, research for the British Arts Festivals Association (Allen and Shaw, 2001) suggests that arts festivals can be grouped into several categories, including high profile general celebrations of the arts, festivals that celebrate a particular location, art form festivals, celebration of work by a community of interest, calendar (including cultural or religious festivals), amateur festivals and commercial music festivals. It could be summarised that arts festivals involve the celebration of a theme or event, of human creative skill in areas such as poetry, painting and music and may involve the celebration of an individual artist, artists or historical art event. These celebrations are held for a limited period of time, annually or less frequently, and are open to the public.

Arts festivals are now a prominent feature of cultural life in Britain – there are estimated to be over 550 annually in the UK, not to mention many more local one-day community festivals and carnivals (Rolfe, 1992; Allen and Shaw, 2001). Local authorities are considered to have been an important factor behind this development as they have sought to capitalise on the tourism and economic development opportunities that festivals potentially bring to a region (Rolfe, 1992).

EVENT EVALUATION

Evaluation is becoming increasingly recognised as a valuable tool in demonstrating success and achievement of objectives. This may be driven by internal management requirements (for example, to evaluate against the objectives, evaluate finance and use of resources, audience satisfaction and aspects of the programme) or external stakeholders (for example, economic, social/cultural and environmental impacts, achievement of audience development objectives).

Festival and event evaluation research has explored a number of areas, including evaluating educational objectives (Gitelson, Kerstetter and Kiernan, 1995), the use of desktop mapping as a marketing tool for event planning and evaluation (Verhoven, Wall and Cottrell, 1998), effectiveness of event leadership training (Tzelepi and Quick, 2002), attendee motivation to attend events (Nicholson and Pearce, 2000; Lee, Lee and Wicks, 2004; Gursoy, Spangenberg and Rutherford, 2006) and why festivals fail (Getz, 2002). One topic that has been given greater attention than perhaps others is the evaluation of (service) quality/customer or attendee satisfaction at various event types (Crompton and Love, 1995; Hultsman, 1998; O'Neill, Getz and Carlsen, 1999; Brewster and Jevons, 2000; Getz, O'Neill and Carlsen, 2001; Bordeau, De Coster and Paradis, 2001; Thrane, 2002; Taylor and Shortland-Webb, 2003; Pechlaner, Helfricht, Raich, Zehrer and Matzler, 2004; Cole and Illum, 2006).

However, to date, event research has tended to focus on, particularly economic, impacts (Gitelson, Kerstetter and Kiernan, 1995; Formica, 1998; Getz, 2000; Harris, Jago, Allen and Huyskens, 2001; Hede, Jago and Deery, 2003), with limited published research on the approaches of individual organisations to other forms of evaluation. In addition, much of the research appears to be from a tourism, rather than an organisation, perspective (Mossberg, 2000). Getz (2005) points out that it is not enough just to consider the economic impacts. He identifies that the social, cultural and environmental effects that can add to the development of society should also be considered. Further, Faulkner (1997) comments that evaluation has a significant internal role to play in the input that evaluation has to the organisation's continuing planning and management processes.

For management, the most basic information required for festival evaluation is the attendance at the event (Getz, 2005).

This view is echoed by Rolfe (1992), who identifies that audience size and attendance are the principal means by which festivals evaluate their success. It may also be useful to have a set of benchmarks, targets or solid reference points in terms of quality, methodology and precedent against which progress can be measured (Mosley, 1999).

In a study commissioned by the former Office of Arts and Libraries on arts festivals in the UK (PSI, 1992) a number of festival organisers were interviewed to ascertain how they measured the success or otherwise of their festival. The main focus of this research was on evaluation methods and did not appear to examine the evaluation process in any great detail. More recent studies by British Arts Festivals Association (BAFA) (Allen and Shaw, 2001, 2002) focused on the economic, social and cultural impacts of festivals – the research was presented as a summary of the state of festivals today but did not focus on the process of evaluation or the methods used.

Many key textbooks in the events management field have explored evaluation to a greater or lesser extent (O'Toole and Mikolaitis, 2002; Goldblatt, 2004; Shone and Parry, 2004; Silvers, 2004; Getz, 2005; Bowdin *et al.*, 2006). Despite evidence that suggests that there is a growing interest in evaluation research in the tourism and events sectors (Mossberg, 2000), there does not appear to be a great deal of research concerning evaluation of arts festivals. This is found particularly to be the case in the UK (Prime, 1998).

There are a wide range of definitions of evaluation in the literature. Getz (2005, p. 378) defines evaluation as, 'the subjective determination of worth – to place a value on something,' while Bowdin *et al.* (2006, p. 413) identify it as, 'The process of critically observing, measuring and monitoring the implementation of an event in order to assess its outcomes accurately.' In the wider arts context, Jackson (2004, p. 8)

identified evaluation succinctly as the 'art of asking interesting and provocative questions.' It is clear that although there are different perspectives on evaluation, it needs to measure what the organisers have identified as required within the process. As a result, thought is required by the organisers to ensure that the metrics used will result in the required data.

Impact Evaluation

Festivals have important economic impacts on local and regional economies as well as producing a range of social and cultural impacts. Yet anecdotal evidence suggests that much of the research produced remains internal to the organisation, with only brief summaries entering the public domain as press releases or reports, rather than into the published research literature. This paper does not seek to explore impacts related literature, as this was beyond the remit of the current study; however, a number of authors have reviewed impact evaluation methods. Burgan and Mules (2001) discussed issues relating to sampling and propose making better use of existing data when undertaking economic impact studies. Carlsen, Getz and Soutar (2001) explored event impact evaluation in Australia and internationally through a survey of event industry experts. They concluded that a standardised model has not been produced, despite there being a clear need for this, and presented a number of pre-event and post-event evaluation criteria that may be adopted. Further work by Fredline, Jago and Deery (2003) developed a research instrument for social impacts, which may form the basis for a standardised approach, while Delamere (2001) and Delamere, Wankel and Hinch (2001) discuss the development of a scale for measuring social impacts of community festivals.

From a practitioner perspective, recently the Arts Council of England has funded a range of projects relating to arts evaluation, which may have application for festivals. Reeves (2002) provides an overview of arts impacts research based on the literature available, focusing on the economic and social aspects. One of the objectives of the study was to develop a practical resource to assist practitioners with effective evaluation. Further work commissioned from Comedia (2003) includes an extensive practical overview of evaluation together with supporting evaluation tools developed in Microsoft Excel – participant and audience questionnaires and a project data template. The idea of the evaluation toolkit has also been explored by other arts organisations within the UK. The Arts Council of Northern Ireland (Annabel Jackson Associates, 2004) focus on the evaluation of social impacts on participants of arts groups working in community and voluntary sectors. Further, the Scottish Arts Council (2003; Dean, Goodlad and Hamilton 2001) developed a pilot online toolkit for people and organisations to evaluate Scottish Arts Council funded arts projects, using the toolkit as condition for grants. These developments are also mirrored internationally – for example, Jackson, Houghton, Russell and Triandos (2005) report on the development of a software based Festivals Do-it-Yourself (DIY) Kit, developed by Arts Victoria in Australia, for event organisers of regional festivals to assess the economic impact of their own events.

Purpose of Evaluation

Evaluation is carried out for a variety of reasons and may be linked to the interests of the group or stakeholder that is pressing for it to take place – this will affect the factors that the evaluation focuses on (Feek, 1988; Bowdin *et al.*, 2006). Evaluation can be distinguished from everyday comment or opinion as it is concerned with making judgements against agreed criteria (Feek, 1988). These criteria should be based on the aim and objectives that an organisation has set, which, as Bowdin *et al.* (2006) suggest, should be specific, measurable, achievable, realistic and time-specific (generally referred to by the acronym SMART) and also link into the mission of the organisation (Phillips, 1993). Finally, Getz (2005) notes that it is important to consider effectiveness and efficiency, i.e. were the actions taken to achieve the aims and objectives effective and were the resources used in doing so used efficiently?

Research indicates that financial viability is crucial to a festival's survival, consequently this is an area that is closely monitored, although this is not generally recognised as a sufficient measure of a festival's success. Bowdin *et al.* (2006) note that one of the main reasons that event managers evaluate is to report to internal and external stakeholders, though other factors of evaluation are considered equally important for a festival to measure its success, for example artistic content and audience attendance (Rolfe, 1992). Many 'smaller' British arts festivals are run by volunteers on low budgets and are purely concerned with 'providing entertainment' and consequently may not have either the budget or, they may believe, the cause, to evaluate their festival. Conducting a full-scale economic impact evaluation can be a costly exercise and may also be unnecessary for many festivals. This perhaps suggests that there are cost limitations to arts festival evaluation but also that the practicalities of evaluation have to be fully considered in the context of the organisation.

Authors have a range of views on the purposes of evaluation. Woolf (2004) focuses on two main purposes, these are to improve practice during a project and for future projects and to show what happened as a result of a project. Jackson (2004) summarises that evaluation is about evidence, causation, different perspectives, reflection and learning. Further, Getz (2005, p. 377) presents a number of practical reasons for evaluation, including the need to: identify and solve problems; find ways to improve management; determine the worth of the event; measure success or failure; identify costs and benefits; identify and measure impacts; satisfy sponsors and authorities (accountability) and gain acceptance, credibility and support.

Although these may be admirable aims, they are not always the primary motivation, with Rolfe (1992, p. 73) noting that, 'Festival organisers evaluate principally when they become aware of problems in the existing programme and feel it necessary to review the festival programme, its marketing or some other aspect of the festival.' However, Getz (2005) goes on to suggest that, rather than being an infrequent task for solving problems or generating new ideas, evaluation directs the marketing and planning functions and enables an organisation to continually learn about itself, its environment and ways to improve its management. Evaluation can be viewed as an extension of the control function of management and assists in the development of management processes and procedures for next hosting the event (Hall, 1997).

Types and Levels of Evaluation

Bowdin *et al.* (2006) and Getz (2005) identify three basic types of evaluation; formative (pre-event assessment), process (monitoring) and outcome or summative evaluations (post-event). Post-event evaluation is generally viewed as the most common type of evaluation, with the visitor survey/questionnaire or the management debrief features of this method. Further, Woolf (2004, p. 51) suggests three levels for evaluation, based on the experience within the organisation and how this will inform management processes, which can be used as the starting point for self-evaluation. These levels are illustrated in Figure 1. Level one suggests that the

organisation is aware of the importance of evaluation, and usually plans to obtain some feedback, but they are not sure how to feed back the findings into future projects.

The organisations carrying out evaluation at level two discuss evaluation at the planning stages of a project and decide how to collect data on their projects. However, they sometimes find it difficult to collate and interpret the data. The organisations carrying out level three evaluations view the process as integral to their practice, using a variety of methods to collect evidence and document projects. In the final level, the organisation views evaluation as assisting them to make decisions and deliver higher quality arts projects.

There are a number of important considerations when deciding on the approach to take to evaluation, including, who will actually do it. Feek (1988) notes that objectivity, competence and appropriateness are important elements to consider when deciding who should carry out an evaluation. The evaluation process either can be the responsibility of a central committee or of a senior manager, however the credibility of the evaluation may depend upon the person in charge as well as the methods used (Getz, 2005). Organisers need to balance bringing in an outsider, which costs money, against using an insider, which would cost the

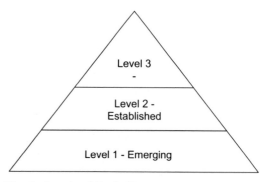

Fig. 1. Levels of Evaluation
Source: Adapted from Woolf (2004, p. 50).

organisation in time and potential credibility of the findings (Feek, 1988).

Process and Methods of Evaluation

With the range of literature available, it is useful to explore the application of a particular approach to evaluation within this study. Woolf's (2004) 'Five Stage Approach to Evaluation' (Figure 2) provides an appropriate model to adopt, as it has been developed specifically with arts organisations in mind and focuses on methods of collecting evidence such as questionnaires, interviews, observation and focus groups, which would be used to conduct audience and participant surveys.

Research suggests that methods of evaluation used by arts festivals vary according to the constitution of the festival, and often take the form of monitoring rather than evaluation (Rolfe, 1992). Mosley (1999) makes the distinction between monitoring and evaluation by suggesting that monitoring is a way of collecting information such as numbers and actions about the progress

of a project at any given time, and refers to monitoring as a 'spot checking process'. Rolfe (1992) also suggests that more formal methods are likely to be used in festivals that are run by local authorities. These may include performance indicators, cost benefit analysis, economic impact studies and meeting 'Best Value' criteria.

No matter which methods are chosen, it is likely that both quantitative and qualitative data will be required to evaluate achievement of the objectives. These may require a variety of tools, depending on the data and evaluator requirements, including questionnaires, visitor surveys, interviews, observations, focus groups and analysis of documents, for example, financial and management records. This paper will identify which of these tools are in use within the sample arts festivals.

METHODOLOGY

The literature review identified that, although evaluation has been discussed in events management texts, and published evaluation research has tended to focus on impacts, there are few published studies that identify how arts festivals evaluate or what evaluation methods are actually used. This exploratory study therefore aims to investigate evaluation practice in arts festivals. Given the range of festivals available and the specific evaluation focus of the study, seven arts festivals were purposively selected. Criteria used to select festivals for inclusion were that they should be a BAFA registered multi-arts festival taking place annually over a period of three days or more; receive some form of government funding and business sponsorship; have registered charity status – be professionally run by paid staff and not solely organised by local authorities. The festivals were from across the UK. These criteria were developed to ensure that only professionally run festivals were selected, which the literature

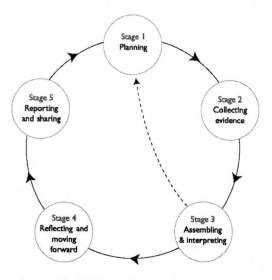

Source: Woolf (2004, p.8)

Fig. 2. Five-stage Approach to Evaluation
Source: Woolf (2004, p. 8)

suggested were more likely to conduct some form of formal evaluation, due in part to the requirements of stakeholders and resources available. Although this small sample does not allow for the results to be generalised to other festival types, or necessarily other arts festivals, it does provide an indication of evaluation in practice.

The festival director was contacted in order to identify the most appropriate person to take part in the research, resulting in the participation of a senior level manager or director with responsibility for the festival overall or the marketing function. Semi-structured telephone interviews were undertaken to allow an in-depth understanding to be developed of the evaluation process, the motives and methods behind arts festival evaluation and to gain an insight into the management of evaluation. Interviews were tape recorded and transcribed to ensure that a reliable record of the interviews was created and then manually analysed and coded, using a content analysis approach, to extract key themes, issues and ideas and to classify results.

RESULTS AND DISCUSSION

Following a review of the literature, and the interview transcripts, a number of key themes emerged. These related to festival aims and objectives, understanding of evaluation, its significance to arts festivals, the purpose, elements to be evaluated, methods, tools and techniques used, the effectiveness of these methods and how evaluation is managed.

Festivals interviewed for this study were asked to provide background information on their festival. Table 1 provides an overview of the festivals included in the study. All of the festivals interviewed are companies limited by guarantee and have charitable status. All festivals consider themselves to be 'mixed arts' festivals.

Festival Aims and Objectives

The literature suggests that in order to be able to evaluate effectively, it is necessary to clearly articulate the overall direction of the festival. This may be expressed in the form of an overall mission or aim, ideally supported by clearly stated objectives. Edinburgh International Festival articulates its overall direction in a mission statement, with the festival aiming 'to be the most exciting, innovative and accessible festival of the performing arts in the world, and thus promote the cultural, educational and economic well-being of the people of Edinburgh and Scotland,' while the overall aim of Harrogate International Festival is 'to continue to achieve an arts festival of international repute.' It is clear that both these festivals have developed international reputations for excellence in the arts and therefore they seek to continue this success by providing measurable objectives (Table 2). Canterbury Festival provided two aims for the festival, the first appearing to focus on the educational purpose of the festival, 'to promote and maintain, improve and advance the education of the public in the arts including: drama, mime, dance, singing, opera and music, ballet and the visual arts generally,' while the second may be seen to articulate the balance that is required between vision and operational necessity in developing the festival, 'we seek to match artistic quality with financial reality, accessibility with innovation and sustenance with growth and development.'

Although all festivals interviewed appeared to have a clear understanding of why they existed, which is articulated in their charity objects, Greenwich and Docklands, Salisbury and Norfolk and Norwich Festivals were unable to supply clearly stated objectives, as at the time of this study they were under review.

As it is difficult to conduct evaluation without any clearly defined criteria to measure against, this perhaps suggests that

Table 1 Background to the festivals

	Company Status	Type	No of Events	Audience size	No of Employees
Brighton Festival	Charitable status, Ltd by Guarantee	Mixed Arts	600–700	430,000 attendances	20 f/t, 20 volunteers
Canterbury Festival	Charitable status, Ltd by Guarantee	Mixed Arts	200–250	75,000	5 f/t, 1 p/t
Edinburgh International Festival	Charitable status, Ltd by Guarantee	Mixed Arts	180	400,000	22 f/t, 3p/t
Greenwich & Docklands Festival	Charitable status, Ltd by Guarantee	Mixed Arts	50–10	55,000	3 f/t, 1 p/t
Harrogate International Festival	Charitable status, Ltd by Guarantee	Mixed Arts	50	32,000	4 f/t, 1 p/t
Norfolk & Norwich Festival	Charitable status, Ltd by Guarantee	Mixed Arts	80	15–20,000	4 f/t
Salisbury Festival	Charitable status, Ltd by Guarantee	Mixed Arts	100	50,000	5 f/t, 5 p/t

some of the festivals interviewed, including Harrogate International, Edinburgh International and Canterbury, may be better placed to conduct effective evaluations than others.

Understanding of Evaluation

One of the key points to emerge was that all of the festivals see evaluation as a way of determining success, by reviewing what has been done against the objectives set (whether these are clearly articulated or not), with the aim of feeding results into the planning process for the future. Canterbury Festival described the evaluation process as a vital, integral part of the year-long administration culminating in a staff debrief and report post festival, which feeds into the planning for forthcoming years. They noted that evaluation is based on not only box office revenue, but also a wide range of qualitative and quantitative factors.

Harrogate International Festival highlighted the educational and quality management value of evaluation, believing it is, 'a way of learning from what you are doing and is a practical way of improving what is being done.' Salisbury festival believed that evaluation involves identifying key areas to work on for the following year and feeding that information into the planning process. Similarly, it was also pointed out by the Greenwich and Docklands Festival that evaluation is learning from strengths and weaknesses so that information can be taken into account in whatever is to be done next. Further, Brighton Festival noted that evaluation concerns a whole range of information, which can be qualitative or quantitative, that is fed into future planning. Evaluation involves acting on outcomes from

Table 2 Examples of festival objectives

Festival	Objectives
Edinburgh International Festival	• Presenting arts of the highest possible international standard to the widest possible audience • Reflecting international culture in presentation to Scottish audiences and reflecting Scottish culture in presentation to international audiences • Presenting events which cannot easily be achieved by any other UK arts organisation through innovative programming and a commitment to new work • Actively ensuring equal opportunities for all sections of the Scottish and wider public to experience and enjoy the Festival. • Encouraging public participation in the arts throughout the year by collaborating with other arts and festival organisations
Harrogate International Festival	• To contribute to the educational, cultural and economic well being of the District and people of Harrogate and Yorkshire. • To promote, maintain and advance education in and knowledge, understanding and appreciation of the arts. • To promote and encourage arts of high standard. • To bring together a programme of events in an innovative way that cannot be easily achieved by other organisations in the region. • To offer equal opportunity for all sections of the public to experience, enjoy and participate in the arts. • To take one year with another to achieve financial break-even. • To maintain a level of financial reserves appropriate of the changing needs of the Festival.

festival reports and taking those on board for the following year and is an informed way of seeing what is successful.

It would appear that there is a clear understanding of evaluation in the context of events management. This is based, on a practical level, on the data requirements of the festivals and feeds into the reporting structures and future planning.

Significance of Evaluation to Arts Festivals

Anecdotal evidence suggests that, although the benefits of evaluation are widely reported in terms of assisting in future planning and quality improvement, events do not always have the time or resources to evaluate effectively. It can, therefore, sometimes be neglected altogether or considered as an afterthought. However, evaluation was reported to be a very positive thing to do. Harrogate International, Edinburgh International and Greenwich and Docklands festivals suggested that evaluation is an essential part of the planning cycle and forms an integral part of the planning process. Canterbury agreed and noted that constant evaluation is critical to continued growth and success of the festival. Norfolk and Norwich Festival suggested that evaluation was important as an acid test to establish how well the festival had done against their budgets by looking at the box office targets and outcomes.

Canterbury Festival noted that evaluation feeds into programming, marketing, audience development and into funding applications. In terms of artistic evaluation,

Brighton Festival believed that evaluation could affect the level of programming, whereas the Norfolk and Norwich Festival commented that beyond financial evaluation it is difficult to measure some aspects as the arts concern subjective elements, such as developing an appreciation of the arts, which are difficult to assess. Harrogate International Festival also believe it is important to be clear about what is being done, why it is being done and how effectively it is being done. A similar issue for the Brighton Festival was that they evaluate event logistics, i.e. what was difficult to manage, what was easy to manage. The Norfolk and Norwich Festival pointed out that they have to evaluate as a requirement of funding, whereas the Harrogate International Festival suggested that to evaluate because an organisation has asked for it to be done is possibly the worst reason to do it, because that may limit the motivation to do it.

It would appear that arts festivals view evaluation as a crucial part of the event planning process; which they believe has to be done in order to avoid repeating mistakes and plan successfully for the next year. It has been suggested that evaluation should be driven by the festivals themselves, linked in to their objectives, and not carried out solely for external purposes and should involve more than purely financial evaluation.

Main Purpose of Evaluation

A number of festivals including Canterbury, Salisbury and Edinburgh International, suggested that their main focus with evaluation concerns audiences, with the Canterbury Festival noting that evaluation informs its audience development plans.

The Harrogate International Festival pointed out that for them evaluation was a learning curve and the main point was to get a grip of where they are, before being able to improve. Similarly, the Salisbury Festival pointed out that the main purpose in

evaluation was to get it right next time. A viewpoint suggested by the Edinburgh International Festival was that evaluation is a complex issue because of the variety of issues that require addressing.

The main purpose to arts festival evaluation, therefore, could be said to depend upon who it is that is requesting the evaluation, while the main focus appears to lean towards audiences.

Elements of the Festivals that are Evaluated

The festivals evaluate a variety of elements, with one of the key elements evaluated by all festivals being finance, perhaps as a result of needing to be accountable to funding bodies, but also for internal management. Salisbury Festival argued that the areas that tend to be most heavily evaluated have a direct relationship between the programme and the audience, whereas the festival tends not to evaluate areas such as venue suitability and front-of-house operations. Norfolk and Norwich Festival measure box office data against targets, whereas the Harrogate International Festival and Edinburgh International Festival measure financial performance against artistic performance. Another key area highlighted was that of marketing and in particular attendance and catchment areas. Marketing was viewed as a key area among all of the festivals interviewed. This goes some way to support the earlier discussion that the principal means by which arts festivals evaluate is by audience size and attendance. However, findings from this study show that arts festivals also focus on other areas, for example public relations, programming, sponsorship, education and staff performance.

Methods, Tools and Techniques

Festivals use a variety of methods, tools and techniques to collect data for their evaluations. Table 3 summarises these.

Table 3 Evaluation methods, tools and techniques

Festival	Data	Technique

Brighton Festival	Audience Research	Face-to-face interviews
		  Web trend analysis
		  Questionnaires

Canterbury Festival	Audience Research	Booking form analysis
		  Focus groups
		  Attendance/sales figures for ticketed and
		  non ticketed events
		  Audience feedback during festival

	Artist/Community	Questionnaires
	Participants Research	Artist feedback
	Financial	Budget monitoring & targets Box office revenue Assessing sponsorship targets
	Media	Print column inches Broadcast local & national
	Staff	Debrief (full time, part time & all volunteer stewards)

	Web Tracking	In the process of developing a new website with tracking
		  statistics for audience analysis and will include a
		  feedback/review form

Edinburgh International Festival	Audience Research	Face-to-face questionnaires
		  Self completion questionnaires
		  Feedback forms
		  Focus groups
		  Observation

	Media	Column inches, impact
	Visiting promoters	Monitor number of national/ international promoters
	Sponsors	Formal evaluation and sponsor objectives Street interviews with public
	Financial	Targets against actual
Greenwich & Docklands Festival	Audience Research	Face-to-face interviewing Questionnaires
	Financial	Budget monitoring
	Operational	Staff debriefing
	Stakeholders	Meetings, discussions
Harrogate International Festival	Audience Research	Focus Groups
	Artistic	Telephone research

	Financial	Benchmarking
		  Financial bottom line
		  BCG Analysis
		  Benchmarking

	Educational	Focus groups feedback forms
	Staff	Appraisals event feedback forms
	Media	Column inches, outcomes

Norfolk & Norwich Festival	Financial Education	Targets against actual budget
		  Video
		  Photographs

Salisbury Festival	Audience Research	Sales figures Marketing report

One of the main methods of evaluation used by arts festivals is to analyse audiences and audience attendance figures. The festivals use a range of techniques to analyse audiences, one of the most common techniques reported was to monitor and evaluate box office figures, which the Edinburgh International Festival suggested was an evaluation at a very basic level.

Edinburgh International Festival was the only festival for which impact studies were known to have been undertaken. They suggested that economic impact evaluation can be useful for organisations such as the Edinburgh International Festival, as it makes their argument for public funding easier. However, it should be considered whose role it is to conduct this sort of evaluation, as it is the people who are providing the festival with public money who are interested in how the money is being spent and the resulting economic impacts. It is, therefore, perhaps more appropriate that they should conduct this type of evaluation. The festival also suggested that to them social and cultural impacts are far more interesting, especially as their mission includes the promotion of cultural and education well-being of people in Edinburgh and Scotland, in addition to economic aspects.

The methods used to establish the success or otherwise of arts festivals vary and in some cases take the form of monitoring rather than evaluation in a true sense. However, it does appear that since earlier studies, all of the festivals have begun to utilise a wider variety of evaluation techniques, such as benchmarking, interviews, focus groups and visitor surveys.

Effectiveness of Evaluation Methods

The festivals were asked how effective the methods of evaluation used were. The general feeling among all festivals was that this was a difficult question to answer. Salisbury Festival suggested that this was because there were so many different methods in use, but they did believe that there was room for improvement. Edinburgh International Festival pointed out that methods used could be more effective and that they are unable to do as much as they would like to, due to the costs involved. They made the point that they do what they can and they are very clear about why they are doing it, what they want the information for and are very clear about 'testing' what they do against their objectives.

This perhaps indicates that evaluations need to be carefully planned and that the festivals need to be dear about why they are inductility the evaluations and what it is that they are trying to find out.

Management of Arts Festival evaluation

Festivals were also asked how the evaluation process was managed. This included establishing how results from evaluations are used, how the results are presented when evaluation takes place and who is responsible for managing the evaluation process.

The results of evaluation had a variety of uses. Evaluation can provide a range of data, which can help inform internal management of the festivals and indicate areas for improvement. Brighton Festival suggested very practical uses, for example, evaluations would show where visitors to the festival are coming from, enabling the festival to address where adverts and brochures are placed. Harrogate International Festival use results in a practical way to assess print and design, which, along with sponsorship, volunteer and staff evaluations, is fed into future planning. However, the more important artistic and financial evaluations, which are difficult to quantify, need to be backed up by further audience research. All of the Festivals pointed out that they use their results to provide information to external sources, including satisfying sponsors, authorities and other stakeholders. Canterbury

Festivals use their evaluations in making grant applications, for marketing purposes and assessing their strategic place on a national level. They consider it important to have statistics, which enable them to compare with statistics compiled by BAFA, to show where they fit into the national picture in terms of how festivals are run and how festivals market themselves. With the availability of this data, it perhaps suggests that statistics at a national level could form benchmarks that festivals could measure against.

Results from evaluations are presented in a variety of formats. The most common form was in a formal report presented to the festival's board of directors and other relevant stakeholders, such as funding bodies and sponsors. This then provides a sound basis for future decisions by ensuring that relevant issues that require attention are identified and dealt with by feeding in to the internal management processes.

An important factor in the evaluation process is when the evaluation takes place. Norfolk and Norwich, Salisbury and Canterbury Festivals reported that they evaluate after the festival, whereas all of the other festivals evaluate throughout the festival as well as post festival, with Edinburgh International Festival suggesting that evaluation is a continuous process. It would appear that festivals follow the various levels of evaluation as discussed earlier. However, it is evident that some caution is required in evaluation during the festivals (sometimes referred to as monitoring), as feedback can be anecdotal and could lead to hasty decisions being made.

Responsibility for managing the evaluation process differed between festivals, with a variety of approaches highlighted. Salisbury Festival pointed out that initially the senior managers, i.e. Finance, Marketing, General Manager and the Festival Director, are responsible for evaluating their areas. Evaluation is then 'spread down', so that feedback is obtained from everybody. Similarly, the Harrogate International Festival meet up as a team, after the festival, and each member of staff is responsible for organising evaluations for their area of the festival. Overall responsibility for evaluation is divided up between the team. Norfolk and Norwich Festival stated that overall responsibility for evaluation lies with the Festival Director, who liaises with the board of Directors and the Marketing Officer. Canterbury Festival noted that all staff (e.g. Director, Marketing Manager, Development Manager, Outreach Co-ordinator, and Festival Administrator) are involved in constant evaluation. Results would suggest that overall responsibility for evaluation lies with festival senior managers and that Marketing Managers have an important role to play in arts festival evaluation. However, it is noted that evaluation may involve the whole festival team and that external assistance in planning and conducting evaluation is useful.

Festivals were also asked if staff within the festivals had received training in evaluation. The overall response from all festivals was that the staff, although not specifically trained in evaluation, had attended arts training courses and had a wealth of experience. They also pointed out that they are aware of publications, such as 'Partnerships for Learning' (Woolf, 2004) and utilise other festivals' and arts organisations' questionnaires as guidelines for developing and producing their own. This demonstrates that the publications available through organisations, such as the Arts Councils, together with the sharing of best practice between festivals, helps in informing the development and improvement of evaluation in arts festivals.

Finally, the festivals were asked if they could identify what resources they allocate to evaluation. This received a mixed response. All of the festivals found it difficult to quantify a figure for evaluation, possibly as many costs are hidden in staff time, apart from where externally funded evaluation

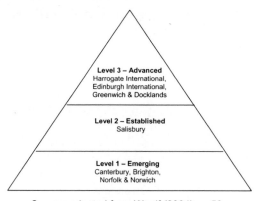

Source: adapted from Woolf (2004), p. 50.

Fig. 3. Levels of Evaluation
Source: adapted from Woolf (2004), p. 50.

takes place. Salisbury Festival suggested that for their marketing team, perhaps a quarter of the year is spent on evaluation, whereas the rest of the team spend much less time, and that they believed that more time should be devoted to evaluation. They indicated that they received a specific grant for market research, which allowed them to conduct some additional evaluation, but they are unable to do this from their core funding. The Harrogate International Festival pointed out that from time to time they utilise external resources, for example a Public Relations consultant, who would look after particular elements of the evaluation process. This is intended as an investment, as the consultant sets up a structure, which the festival could run in future years.

There appears to be no single approach to funding evaluation and festivals believe that they are unable to do as much as they would like due to lack of resources. This reiterates the point that perhaps funding bodies could be doing more to assist in evaluating festivals.

CONCLUSION

This paper aimed to explore art festival evaluation practice through reviewing the literature base and presenting the findings of an exploratory study of evaluation practice within seven arts festivals. The literature review revealed a paucity of previous festival evaluation research, particularly in relation to approaches by festivals themselves, and found that the focus of much of the event evaluation literature tended towards the economic impacts.

This exploratory study has revealed that evaluation is considered a crucial part of the planning process for arts festivals and has become increasingly important in recent years for some festivals, such as the Canterbury, Harrogate International, Edinburgh International and Greenwich and Docklands Festivals. This is perhaps due to festivals being required to justify their use of public funding and demonstrate their accountability to sponsors and other key stakeholders, but also due to developments in best practice in festival management. One of the most important reasons for event evaluation was thought to be for an organisation to learn and adapt and to continually improve quality and encourage innovation. Based on a review of the literature, findings from the research, and the characteristics of Woolf's (2004) criteria, it is possible to illustrate the levels of evaluation apparent within the sample festivals (Figure 3).

The effectiveness of methods used was found to be difficult to determine, due to the range of methods used and the subjectivity of arts appreciation. The Harrogate International Festival highlighted the difficulties in their use of artistic and financial benchmarks and it was suggested that perhaps national benchmarks could be established. All festivals believe that they could do more evaluation, which would assist them in measuring whether their aims and objectives have been achieved for internal management purposes but also provide evidence for external stakeholders. Within the study, economic impact evaluations of the Edinburgh International Festival highlighted the benefits to the local economies and made it easier to gain further funding. However, it is suggested that it can be dangerous for festivals to go down the economic impact evaluation route alone, especially if the festivals' aim and objectives are concerned with social and cultural issues rather than economic impacts, and, therefore, if impact studies are considered, they need to consider economic, social, cultural and environmental impacts.

The study has shown that arts festivals carry out evaluation at varying levels of advancement and effectiveness and use a wider range of techniques to evaluate, than earlier research suggests. This is perhaps linked to the increasing importance placed on evaluation by some festivals, the arts sector in general and by the various stakeholders. In order for arts festivals to plan successfully and to justify their use of public funding, evaluation must be integral to the planning process, with clear aims and objectives stated to enable effective evaluation. The range of evaluation toolkits and publications available are providing the basis for professional development and for standardised approaches to be developed. Although the purpose and measurements will differ depending on the individual festival's requirements, these tools assist busy festivals in providing a tried and tested approach and, should they wish, ensuring consistency and comparability of methodology. Organisations, such as Arts Council of England and the British Arts Festivals Association, and other interested stakeholders, such as local authorities, could provide further encouragement for evaluation through assistance with funding and training, while sharing knowledge among the festivals would assist in achieving a standardised approach. Through sharing best practice, and developing the robustness and design of evaluation in the future, arts festivals will be able to demonstrate the value of all aspects of their programme and increasingly the economic, social and environmental impacts.

The festival sector is fast moving, with an increasing range of evaluation material being introduced over the past few years. The festivals in this study continue to develop in terms of size and scope, together with improvements in their management approaches, systems and procedures. This study has provided an insight into evaluation practice in seven arts festivals within the UK. However, further research is required in order to investigate evaluation practice in a wider sample of arts festivals, in other sizes and types of festivals and events, and to fully understand the effectiveness of evaluation.

ACKNOWLEDGEMENTS

The authors would like to thank the participating festivals for contributing their time and experience of evaluation to the study and for permission to include their details. Grateful thanks are also due to the anonymous paper reviewers for their constructive comments and suggestions.

REFERENCES

Allen, K. and Shaw, P. (2001) *Festivals Mean Business: The Shape of Arts Festivals in the UK*, London, British Arts Festivals Association.

Allen, K. and Shaw, P. (2002) *Festivals Mean Business II: The Shape of Arts Festivals in the UK Update*, London, British Arts Festivals Association.

Bordeau, L., De Coster, L. and Paradis, S. (2001) Measuring satisfaction among festivalgoers: differences between tourists and residents as visitors to a music festival in an urban environment, *International Journal of Arts Management*, **3**(1), 40–50.

Bowdin, G., Allen, J., O'Toole, W., Harris, R. and McDonnell, I. (2006) *Events Management, edition*, Oxford, Elsevier Butterworth-Heinemann.

Brewster, J. and Jevons, C. (2000) Event marketing as an antecedent determinant of consumer decision making: the effectiveness of a University Open Day, in Proceedings of the Australian and New Zealand Marketing Association Conference, *Visionary Marketing for the 21st Century: Facing the Challenge*, Brisbane, December, pp. 138–142.

Burgan, B. and Mules, T. (2001) Sampling frame issues in identifying event-related expenditure, *Event Management*, **6**(4), 223–230.

Carlsen, J., Getz, D. and Soutar, G. (2001) Event evaluation research, *Event Management*, **6**(4), 247–257.

Cole, S. T. and Illum, S. F. (2006) Examining the mediating role of festival visitors' satisfaction in the relationship between service quality and behavioral intentions, *Journal of Vacation Marketing*, **12**(2), 160–173.

Comedia (2003) *National Arts Information Project: Evaluation Toolkit*, London, Arts Council of England.

Crompton, J. L. and Love, L. L. (1995) The predictive validity of alternative approaches to evaluating quality of a festival, *Journal of Travel Research*, **34**(1), 11–24.

Dean, J., Goodlad, R. and Hamilton, C. (2001) *Toolkit for Evaluating Arts Projects in Social Inclusion Areas: A Report to the Scottish Arts Council*, Edinburgh, Scottish Arts Council.

Delamere, T. A. (2001) Development of a scale to measure resident attitudes toward the social impacts of community festivals, part II: verification of the scale, *Event Management*, **7**(1), 25–38.

Delamere, T. A., Wankel, L. M. and Hinch, T. D. (2001) Development of a scale to measure resident attitudes toward the social impacts of community festivals, part I: item generation and purification of the measure, *Event Management*, **7**(1), 11–24.

Faulkner, B. (1997) A model for the evaluation of national tourism destination marketing programs, *Journal of Travel Research*, **35**(3), 23–32.

Feek, W. (1988) *Working Effectively: A Guide To Evaluation Techniques*, London, Bedford Square Press.

Formica, S. (1998) The development of festivals and special events studies, *Festival Management and Event Tourism*, **5**(3), 131–137.

Fredline, L., Jago, L. and Deery, M. (2003) The development of a generic scale to measure the social impacts of events, *Event Management*, **8**(1), 23–38.

Getz, D. (2000) Defining the field of event management, *Event Management*, **6**(1), 1–4.

Getz, D. (2002) Why festivals fail, *Event Management*, **7**(4), 209–220.

Getz, D. (2005) *Event Management and Event Tourism*, edition, New York, Cognizant Communication Corporation.

Getz, D., O'Neill, M. and Carlsen, J. (2001) Service quality evaluation at events through service mapping, *Journal of Travel Research*, **39**(4), 380–390.

Gitelson, R., Kerstetter, D. and Kiernan, N. (1995) Evaluating the educational objectives of a short term event, *Festival Management and Event Tourism*, **3**(1), 9–14.

Goldblatt, J. J. (2004) Special Events, edition, Hoboken, John Wiley & Sons.

Gursoy, D., Spangenberg, E. R. and Rutherford, D. G. (2006) The hedonic and utilitarian dimensions of attendees' attitudes toward festivals, *Journal of Hospitality and Tourism Research*, **30**(3), 279–294.

Hall, C. M. (1997) *Hallmark Tourist Events – Impacts, Management and Planning*, London, Wiley.

Harris, R., Jago, L., Allen, J. and Huyskens, M. (2001) Towards an Australian event research agenda: first steps, *Event Management*, **6**(4), 213–222.

Hede, A.-M., Jago, L. and Deery, M. (2003) Special event research 1990–2002: a research agenda, *Journal of Tourism and Hospitality Management*, **10**(3), 1–14.

Hultsman, W. Z. (1998) The multi-day, competitive leisure event: Examining satisfaction over time, *Journal of Leisure Research*, **30**(4), 472–497.

Jackson, A. (2004) *Evaluation Toolkit for the Voluntary and Community Arts in Northern Ireland*, Bath, Annabel Jackson Associates.

Jackson, J., Houghton, M., Russell, R. and Triandos, P. (2005) Innovations in measuring economic impacts of regional festivals: a do-it-yourself kit, *Journal of Travel Research*, **43**(4), 360–367.

Lee, C. K., Lee, Y. K. and Wicks, B. E. (2004) Segmentation of festival motivation by nationality and satisfaction. *Tourism Management*, **25**(1), 61–70.

Mossberg, L. (ed.) (2000) *Evaluation of Events: Scandinavian Experiences*, New York, Cognizant Communication Corporation.

Nicholson, R. and Pearce, P. L. (2000) Who goes to events: a comparative analysis of the profile characteristics of visitors to four South Island events in New Zealand, *Journal of Vacation Marketing*, **6**(3), 236–253.

O'Neill, M., Getz, D. and Carlsen, J. (1999) Evaluation of service quality at events: The 1998 Coca-Cola Masters surfing event at Margaret River, Western Australia, *Managing Service Quality*, **9**(3), 158–166.

O'Toole, W. and Mikolaitis, B. (2002) *Corporate Event Project Management*, Hoboken, Wiley.

Pechlaner, H., Helfricht, M., Raich, F., Zehrer, A. and Matzler, K. (2004) Customer satisfaction with winter sports events: The case of Biathlon World Cup 2003. *Journal of Hospitality and Tourism Management*, **10**(1), 15–25.

Prime, S. (1998) *South East Arts: Festivals Strategy*, Surrey, South East Regional Arts Board.

Policy Studies Institute (PSI) (1992) Arts Festivals. *Cultural Trends*, 15, London, Policy Studies Institute.

Reeves, M. (2002) *Measuring the Economic and Social Impact of the Arts*, London, Arts Council of England.

Rolfe, H. (1992) *Arts Festivals in the UK*, London, Policy Studies Institute.

Scottish Arts Council (2003) *Evaluation Toolkit: The Scottish Arts Council E-Tool for Evaluation*, Edinburgh, Scottish Arts Council, [Internet]

available online at: www.evaluationforall. org.uk, accessed 31 January 2006.

Smith, L. J. (1990) *Dictionary of Concepts in Recreation and Leisure Studies*, Westport, Greenwood Press.

SQW Limited and TNS (2005) *Edinburgh Festivals 2004–2005 Economic Impact Survey Stage 1 Results*, Edinburgh, The City of Edinburgh Council, Scottish Enterprise Edinburgh and Lothian, EventScotland and VisitScotland.

Taylor, R. and Shortland-Webb, G. A. (2003). A delegate evaluation of conference satisfaction, in: K. Weber (ed.) *Advances in Convention, Exhibition and Event Research*, *Proceedings of the Convention & Expo Summit 2003*, Hong Kong, The Hong Kong Polytechnic University, pp. 112–119.

Thrane, C. (2002) Music quality, satisfaction, and behavioral intentions within a jazz festival context. *Event Management*, **7**(3), 143–150.

Tzelepi, M. and Quick, S. P. (2002) The Sydney Organising Committee for the Olympic Games (SOCOG) "Event Leadership" training course – an effectiveness evaluation, *Event Management*, **7**(4), 245–258.

Verhoven, P. J., Wall, D. L. and Cottrell, S. (1998) Application of desktop mapping as a marketing tool for special events planning and evaluation: a case study of the Newport News Celebration in Lights, *Festival Management and Event Tourism*, **5**(3), 123–130.

Woolf, F. (2004) *Partnerships for Learning: A Guide to Evaluating Arts Education Projects*, London, Arts Council of England.

Conceptual paper: An exploration of time and its management for sport event managers

Paul Emery and Alexandru Radu

In practice, the notion of time has largely been taken for granted and in the event management industry it is often regarded as the mismanaged resource (Goldblatt, 2005). In essence, the aim of this research is to conceptually appraise an event manager's use of time and begin the process of developing a user-friendly and non-invasive methodology of time management analysis from which the event industry can benefit.

To further understand what the sports event manager actually does in the implementation phase of an event, this paper appraises the time management practice of one event manager at two professional basketball matches. It applies Hedaa and Törnroos's (2002) two dimensional time – space model, which includes both a chronological approach (the dominant approach of absolute time analysis) as well as integrates the virtually unused approach and concept of kairology – a theory of timing that recognizes performance outcome in context, whereby independent actors and activities need to engage in a shared time-space framework for successful outcomes to be realized.

Utilizing a repeated measures research design, this exploratory research involved two phases of data collection at each event. The first entailed direct observation of the event manager for a full working day at a basketball event. Data on the manager's activity and time usage was collected by shadowing and recording all of the manager's activities immediately before, during and after the match. The second phase constituted a face-to-face, post-event semi-structured interview with the manager. This was used to determine the normality of the event experience, the perceptions and context of the actual time usage, the focus on improvements as well as to identify potential limitations in the data collection methods. Researcher insights were additionally recorded to provide a further perspective of appraising the user friendly nature of both the in situ data collection and time analysis techniques.

Combining the quantitative and qualitative data collection findings, this research provides both a theoretical framework and practical process for further analysis of improving personal and organizational productivity levels.

STUDY BACKGROUND

Managing a successful sports event is said to be dependent upon the effective micro-management of the inter-related components of quality, cost and time (Westerbeek, Smith, and Turner *et al.*, 2005). While the management of quality, finance and human resource (costs) have received considerable attention in the academic and professional literature, the analysis of time has not. This is quite surprising since time is the only perishable resource that cannot be saved, bought or stopped, making it the most highly valued yet wasting asset available to management (Harvard Business School, 2005).

In the complex multi-stakeholder environment of managing a major sport event, time often becomes the project's primary limiting factor, with management frequently possessing just one opportunity to get it right (Emery, 2003). However, in sport event practice, effective time management is rarely analysed, achieved or appraised. Murray (1995) for example cites the Atlanta Olympic Committee staff who on an average spent 14 hours a day preparing for this once-in-a-lifetime event. More recently, Phillips (2005) makes reference to the 2006 Commonwealth Games Chief Executive who, having worked 15 hour a day for many months, was ordered to take a holiday because of concerns for his health. Excessive time usage is not uncommon in the event industry and Goldblatt (2005) reports that an event management career is now renowned for its very long-work hours as well as high degrees of burn-out. Clearly, this scenario is unsustainable in the emerging event profession and provides the primary rationale behind this exploratory research of time management practice.

A review of the research literature reveals that there is little known about what sport event managers actually do, let alone how they actually use their time. In addition, there appears to be no consistent methodology of collecting time related data, nor any established benchmarks of practice. To begin the process of understanding and improving time management practices, this particular case study aims to

- appraise an event manager's usage of time in the implementation phase of a sports event.
- determine a user-friendly and non-invasive method of time related data collection and analysis.

LITERATURE REVIEW

The origins of time management studies can be traced to the early 1900's under the scientific management pioneers of Taylor and Fayol (Cole, 1993). Focussing upon time and motion studies their ideas were based upon the premise that time is money and that a less than optimal use of time meant an organization was, in effect, wasting money. Similarly, Godefroy and Clark (1990) argued that time is distributed equally and democratically to everyone, and its inequality at the organization and personal level can only be attributed to its mis-management and its waste. More recently Adam, Whipp and Sabelis (2002), Odih and Knights (2002) and Lee and Liebenau (2002) have challenged these underpinning assumptions, concluding that time is still a relatively 'unexplored and unquestioned concept' (Odih and Knights, 2002).

Adam *et al.* (2002, p. 16) question the notion that time is money.

> Time passes outside our control while money can be consumed at intentional pace or it can be left to grow. For people the accumulation of days and years means ageing, growing older and therefore closer to death, while the accumulation of money means growth of wealth.... Whilst money stands in a direct quantitative relation to value – the more the better – this is clearly not the case with time, otherwise the time

of prisoners would have to be accorded maximum value.

The concepts of time and money may well possess many similarities, *e.g.*, measurement irrespective of location, context and emotion; however, there are also significant differences between the two. Unlike the management of money, time is not a cumulative resource but a very perishable one that cannot simply be borrowed or saved for use on another day.

Despite these conceptual differences, Adam *et al.* (2002) and others (Odih and Knights, 2002; Hedaa, and Törnroos, 2002) conclude that the assumption that time is money appears to dominate management theory and practice today. This has resulted in time being appraised from a narrow closed system of analysis. This focus is perhaps understandable as global work and lifestyle practices are commonly based upon a monochronic and quantitative approach towards the measurement of time (Lee and Liebenau, 2002). Historically based upon nature's diurnal, seasonal and birth/death lifecycles, most developed cultures relate to a world calendar, where clock time is measured in absolute seconds, minutes, days and years. Similarly, in the work place, increased labour productivity (work output per unit of time) and accountability are achieved through the use of many performance/time related practices. These include time focussed strategies, objectives, performance indicators, industry benchmarks, just-in-time management and project management practices, which are increasingly being coordinated through manual and electronic personal organizers.

With the significant advancement in global technologies, the linear nature of time is becoming even more complex as 24/7 business and pleasure activities become practical realities. Despite such increasing complexity, it appears that every society embodies a notion of a past, present and future absolute, linear clock time. In this sense, it has become a defining feature of human culture and everyday existence (Adam *et al.*, 2002). Since, clock time is the dominant measure through which social relations and structures are currently established, any study on effective time management cannot ignore it.

Hedaa and Törnroos (2002), on the other hand, propose another contemporary and complementary perspective towards the effective management of time, namely that of kairos. It is a combination of the two, Chronos and Kairos as illustrated in Figure 1, which will provide the conceptual analysis of this investigation.

The first approach (chronos) relates to the linear and dominant chronological method of time management study already discussed. It assumes a relatively simple and predictable environment of order and stability as it applies a closed system of quantitative time analysis. Derived from both scientific and human relations management theories, effective time management assumes capable and motivated staff being in control of simple and often repetitive tasks. This type of environment may be encountered in a factory scenario or at an established sport/leisure facility, where performance often entails largely autonomous managers (actors), implementing numerous organizational routines, through standard operating procedures and practices (Bunce, 2000).

By contrast, the concept of kairos denotes 'the theory of appropriate timing for action in differentiated managerial situations in context' (Hedaa and Törnroos, 2002, p. 31). It encompasses a broader consideration of the temporal features of timescape and draws heavily upon contingency/situation management theories. In this sense, the world of kairos may be classified as an open system approach and can be considered a descriptive analytical theory, where management performance operates

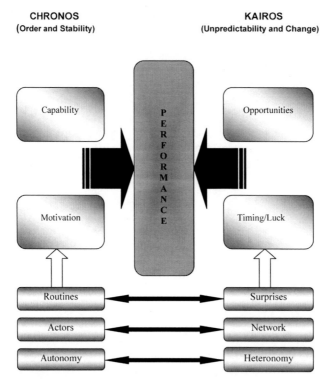

CHRONOS
(Order and Stability)

KAIROS
(Unpredictability and Change)

Fig. 1. Performance Criteria in a Two – Dimensional Time-Space
Source: Adapted from Hedaa, and Törnroos, 2002.

in an unpredictable environment of opportunity structures and good or bad timing or luck. Elaborating further, Hedaa and Törnroos (2002, p. 31) suggest that this involves a management environment of 'heteronomous actors in a world of complexity and surprises embedded in a network of interdependencies.' In other words, for a desired outcome to take place in an uncertain world, management need to look for the appropriate opportunity (the time) to place people in relevant external and internal situations together (network of interdependencies) in an appropriate environment (temporal time-space). On the one hand, chronos appears to explain time usage in terms of a closed system dependent upon predictability, routines and autonomous actors, and at the opposite extreme, the kairos approach assumes

unpredictability, where surprises, networks and heteronomy are the dominant forces (Hedaa and Törnroos, 2002).

To better understand the two different yet inter-related approaches to time management reference will be made to a practical sports example, namely managing the activity of scoring a goal from a direct free kick (the desired performance outcome) in a game of football. To prepare for this eventuality one or more players (the actors) of the team must have the capability and motivation to achieve the task on a routine and frequent basis. From a management perspective this might be considered a relatively closed skill (the same or very similar scenario encountered in many different games) with the successful outcome being largely dependent upon training the individual player(s) taking the free kick. This might explain why players such as David Beckham and Jonny Wilkinson (kicking a goal in rugby union) spend many hours dedicated to autonomously practicing these closed skills so that eventually taking of free kicks become a perfected and highly specialized routine. Equally, they are hugely rewarded both by status and financially, if the outcome is regularly executed in the competitive setting. This management approach would be considered traditional, representing the world of chronos, where the emphasis is focussed on individual autonomous actors implementing controlled routines to achieve success.

In comparison, the world of kairos assumes an environment of unpredictability and change. In the sporting scenario it initially recognizes that the desired performance outcome can never be truly recognized as a closed skill or predictable outcome. Whether or not the player actually scores may be influenced by the opposition, referee, crowd, pitch conditions or even other unpredictable externalities (*e.g.*, a gust of wind).

Kairos, therefore, assumes an open systems environment where each actor or network can potentially influence, positively and negatively, the nature of the outcome. In this sense, desired outcomes are seen to be determined in their simplest form by the presence of at least two (but usually many more) independent streams of events – the ball being projected towards the opponents goal and the opponents not being able to stop it. While the manager may not be able to fully predict or control the exact nature of the outcome, she or he can try to make sure that the necessary range of diverse and interdependent actors (heteronomy) can strategically and collectively (network) coincide to be in the most appropriate place at the right time (temporal timespace). This does not just apply at the absolute moment of taking the free kick, but also before as well as after its execution.

This has meant that football managers try to influence the opportunity structure legally by applying team tactics to pressurize/obstruct the goalkeeper as well as being the first to react to largely unpredictable rebounds from the initial free kick or illegally (*e.g.*, players trying to deceive the referee by *diving* at the 2006 World Cup; or in the 2006 Italian Series A league by influencing the referee selection process). Kairology thus recognizes that no individual actor can fully control the interactive nature of independent actors and events. But management can increase or decrease the likelihood of them occurring by improving the opportunity structure (the chance that more free kicks will be awarded) and the timing of activities (synchronicity of compatible events occurring).

The broader implications to managers operating in a fast moving, turbulent and uncertain global business environment, are that they need to not only manage increasingly complex linear time (chronos) but also take a more holistic approach to ensure that relevant actors are in the right place at the right time (kairos). No more is this apparent than in the event industry, which is characterized by its finite nature as well as its inherent uniqueness and accompanying uncertainties (Watt, 1994; Getz, 1997; Torkildsen, 1999; Emery, 2003). In such an environment, the manager aims to plan and control a perishable service quality usually against an absolute deadline. Possessing a definite start and finish date the importance of managing time in the event industry is clearly highlighted by Torkildsen (1999, p. 471);

> The event, unlike the normal ongoing programme is speeded up and delivered within a short space of time; this concentrates all the advanced planning and actions into specific hours and moments.

Similarly, Goldblatt (2005) supports the notion that event managers must 'manage most minute segments' (p. 59), ~~borates~~ that the entertainment ~~~ess~~ must increasingly consider the ~~~cept~~ of timing, both in terms of how much time the manager has to act or react to situations, as well as how to create memorable special effects. For these reasons, he asserts that time management should now be considered as one of the most essential pillars of successful event management.

Emery (2003) argues further that managing sports events can be about managing some of the most complex events imaginable as they frequently encounter extreme levels of uncertainty and complexity. This is often due to their highly uncontrollable environment of potentially illogical emotional spectator fervour, intense live media coverage, unscripted performance outcome, temporality of host organization existence, and a diversity of stakeholders unparalleled in many other event management scenarios (Emery, 2003). Regardless of the complexities encountered, the reality is that the sports event industry is still affected negatively by the many examples of mismanagement and at worse fatalities (Emery, 2003) simply because chronological time has either run out or conversely because managements have just not reacted quickly enough. Elliott and Smith (1993) and Hartley (1997), for example, provide *post-hoc* reviews of specific football stadia disasters, and independently talk of crisis incubation periods, where a single action has very quickly escalated into crisis inertia and eventually a disaster. In all cases, they suggest that if these individual triggers could have been identified earlier then the subsequent event disasters would not have taken place.

Given this industry scenario as well as the increasing stress being encountered in modern day life (Purser, 2002; O'Brien, 2006), some have argued that 'Time management is [now] a skill that is central not only to success but also [personal] survival' (Phillips, 2002, p. 1). Since it directly impacts workplace productivity as well as everyone's life, it is argued here that time management now warrants considerably greater attention than it has received in the past.

In summary, historically the study of time has been

> Firmly embedded in linear representations of time, qualitative variations that precede their quantified representation tend to be disregarded.
>
> (Odih and Knights, 2002, p. 75)

This has led to a largely prescriptive account of chronological time management in training workshops and perhaps few practical improvements in the event industry over the last decade. In practice, a broader and more open system of contemporary analysis is required, particularly in the context of the implementation phase of a sport event environment. In this context activities and decisions are often made in seconds and minutes, and are responses to a complex array of simultaneous networked communication systems (Emery, 1997).

This research, therefore, aims to determine a more conceptual and contextual understanding of time usage in a technological world of instantaneous communication. In essence, this study endeavours to move beyond the established linear and clock representation of time, and establish an exploratory user-friendly and non-invasive methodology of time management practice for the sport event industry.

METHODOLOGY

Given the need to collect both rich qualitative (context specific) and quantitative (linear time) data the methodological approach chosen was an in-depth case study. Such an approach permits a detailed audit of current practice, whereby positive and negative time usage can be determined (chronos) and the potential influence/cause(s), particularly of negative behaviour, can be retrospectively appraised through identification of the interactive actors pertaining to particular moments in time (kairos).

The sports event case study was chosen on a geographical and access basis. Of two professional sports events available in the researchers' locality at the time of the investigation, the sport of basketball was randomly selected through the toss of a coin. Adopting a test–retest methodology an intensive data collection of the time management practices of the professional basketball event manager was carried out on two separate occasions (Event 1 and Event 2). The rationale for selecting two events was to identify whether the linear time related data collection was purely unique and event specific, or whether consistencies existed over time. In this instance separate seasons were used to ensure that the two home events were completely independent activities from which there would be limited opportunities for related time management learning.

Two different methods of data collection were used at each event. The first involved the direct observation of the manager for every minute of a full working day at the two professional basketball matches. This entailed shadowing and recording (using Dictaphone as well as pen and paper) all manager activities/actions and time usage performed on the match day, *i.e.*, immediately before, during and after the basketball game. The reasoning behind the match day timeslot selection was that the sports event implementation phase was determined, by Emery (1997), to be the busiest and most chaotic of the event life cycle – the moment of truth where time had run out and where the most complex array of stakeholder scenarios interacted. The purpose of this method of data collection was, therefore, to provide an accurate and very detailed chronological account of linear time usage by the event manager.

The second data collection method, one week after the event, constituted a face-to-face, post-event semi-structured interview with the manager. This was to link the specific context (*e.g.*, actor involvement, task unpredictability/importance/urgency and timing – kairology) with the chronos data collection via retrospective analysis. More specifically, this entailed a sequential process that comprised of two activities.

Pre-interview activity

The Manager was provided with the original tabulated chronological summary (worksheet example provided in Table 2), and asked to view and complete his/her perceived low/medium/high rating against each task. This was categorised according to the task importance, urgency, the degree to which the event manager reacted to it, and whether in hind sight it could be a delegated task.

Interview

An interview guide was used based upon four key themes, namely (1) the normality of the event experience; (2) the overall perceptions and explanation of the actual time used; (3) the specific identification of high/medium reactive (surprise), low importance and low urgency tasks (poor productivity) each followed by an analysis of who was specifically involved (actors/networks), the cause of the actual behaviours (heteronomy and timing) and the actual/preferred outcome and finally, (4) feedback on the potential limitations and improvements in the data collection methods.

The second event post-event interview followed a similar format but, in addition, included a detailed appraisal of where and how personal productivity as well as timing could be improved.

To further appraise the methodological process, the researcher also wrote down his insights (positive and negative) regarding the user-friendly nature of the data collection. This was carried out within three days of each data collection period and also repeated immediately after the data analysis phase had been completed.

FINDINGS AND DISCUSSION

Quantitative Analysis

Chronological analysis of the event manager's activity at the first event (Event 1) revealed a working day of 11 hours 30 minutes with the game itself lasting 1 hour 49 minutes. This constituted involvement in more than 150 separate activities that varied in duration from between 1 and 27 minutes. While these were considered to be separate tasks (on the basis that the researcher perceived a new task had actually started) they were not necessarily discretely different in nature. For example, *Preparing the Match Day Announcements, writing cheques* and *supervising the Game* were not undertaken as one-off tasks but were completed as fragmented tasks usually within a certain time frame and triggered by other actions (more details to follow under the Qualitative findings section).

As highlighted in Figure 2 tasks were categorized according to the original nature of the activity, and the five most time consuming activity groupings (cumulatively) were *Supervising/Watching the Game* (119 minutes; 17% of total activity), *Working on the Computer/Typing* (75 minutes; 11%), *Talking on the Phone* (57 minutes; 8%), *Emails* (57minutes; 8%) and *Talking – Disseminating Information* (52 minutes; 8%), respectively. Regarding *Watching the Game* activity, this was not to suggest that the event manager is just like any other passive spectator viewing the match. The reality is that this was a public presence supervisory

role whereby he was in a continuous state of response to incidents either in person or through technological communications.

To create more meaningful and potentially comparable data, these 21 activity groupings were compressed to four activity clusters, namely *Match Day Programme* (43 minutes), *Management Functions/Activities* (259 minutes), *Communications* [subdivided into *Verbal Communications* (220 minutes), *Electronic Communications* (132 minutes) and *Written Communications and Reading* (8 minutes], and *Other Activities* (28 minutes). [See Table 1]

While these four clusters were not mutually exclusive, *Written Communications* for example were clearly involved in the production of the *Match Day Programme*, they were determined on a hierarchical basis. At the highest level were 'event specific tasks' (*e.g.*, match day programme), because this was the terminology and orientation the manager demonstrated towards organizing his time. To further increase the application of the findings to other management scenarios, many of these 21 activity groupings could be seen to be directly related to Lussier and Kimball's (2004) more generic 'management functions/activities' (*e.g.*, resource allocation, problem solving, monitoring) so this formed the second level of the hierarchical compression. Where the remaining activities could not easily be classified under these higher order clusters, they were included in the next level 'communications' (*e.g.*, casual talking), or in the lowest classification level of 'other activities' (*e.g.*, physiological tasks such as eating or going to the toilet). Given the amount of management time spent on *Communication* activities, these were further classified into individual activity clusters of *Verbal, Electronic* and *Written/Reading Communications*. Furthermore, the richness of the data collected permitted additional classification detail as was the case with *Verbal Communications* where, for example, it was identified

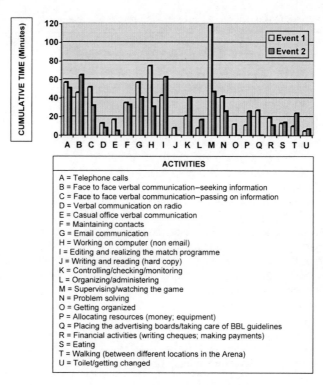

Fig. 2. Comparative Time Activity (Events 1 & 2), Chronos Approach

that Event 1 telephone calls constituted 57 minutes in total – subdivided into incoming (22 minutes; 14 calls) and outgoing calls (35 minutes; 10 calls).

Using the same comparative framework for Event 2 and also referring to Table 1 the findings reveal a similar working day in excess of 10 hours (615 minutes). On this occasion, 197 separate activities were engaged with, ranging from 1 minute activities through to 73 minutes (98% of these activities were less than 15 minute activities in duration). The five most time consuming tasks/activities (cumulatively) at this second event were Proactive Networking (73 minutes; 12%), *Talking – Seeking Information* (65 minutes; 11%), Match Day Programme (63 minutes; 10%), *Talking on the Phone* (51 minutes; 8%) and *Supervising/Watching the Game* (47 minutes; 8%), respectively.

As evident from Table 1, both events despite being on different days of the week entailed very similar management activity clusters and percentage time usage profiles. For example, *Management Functions/Activities* (Event 1 – 38% of total time; Event 2 – 39%) and *Verbal Communications* (Event 1 – 32%; Event 2 – 32%) were virtually the same and the largest percentage difference between clusters was 7% (*Electronic Communications*). Furthermore, even some of the individual activities were remarkably similar in total percentage timings; such as with *Solving Problems* (Event 1 – 6%; Event 2 – 4% of total time), *Financial Activities* (Event 1 – 3%: Event 2 – 2%) and *Talking on the Phone* (Event 1 – 8%: Event 2 – 8%). The only individual activity differences apparent (>7% difference between the two) was an increase in *Proactive Networking* (Event 1 – 0%; Event 2 - 12%) at the cost of

Table 1 Comparative time clusters (events 1 and 2) ~ chronos approach

Level	Activity cluster	Total Minutes (%)	
		Event 1	Event 2
1	**Match Day Programme**	43 (6%)	63 (10%)
2	**Management Functions/Activities**	259 (38%)	241 (39%)
	a) Controlling/checking/monitoring	21	41
	b) Organizing/administering	8	17
	c) Supervising/watching the game	119	47
	d) Solving problems	42	26
	e) Getting organized	12	0
	f) Financial activities (writing cheques, making payments)	19	11
	g) Allocating resources and delegating	11	26
	h) Placing the advertising boards/taking care of BBL guidelines	27	0
	i) Proactive networking (VIP and Guest Room)	0	73
3a	**Verbal Communications**	220 (32%)	194 (32%)
	a) Talking on the phone	57	51
	b) Talking/speaking		
	– seeking information	46	65
	– disseminating information	52	32
	c) Talking on radio before/during the game	13	8
	d) Casual office conversation	17	5
	e) Talking/maintaining established contacts (including outside contacts)	35	33
3b	**Electronic Communications**	132 (19%)	72 (12%)
	a) Emails	57	41
	b) Working/typing on the computer	75	31
3c	**Written Communications & Reading**	8 (1%)	0 (0%)
4	**Other Activities**	28 (4%)	45 (7%)
	a) Eating	13	14
	b) Walking (within the arena)	10	24
	c) Toilet/getting changed	5	7
	Total Working Day	690	615

reducing *Supervising/Watching the Game* activities (Event 1 – 17%; Event 2 – 8%).

As exploratory research with no comparable data available, these descriptive findings begin the process of identifying what an event manager actually does (in this case on a basketball match day) and merely describe how this particular manager used his time on two separate occasions. Similarly, this case study portrayed a very linear absolute time usage, as the practice of multi-tasking was barely evident. The only multi-tasking apparent was when the manager was making telephone calls or in radio communication. Given that these tasks were usually less than 5 minutes and the primary focus was the person who was being spoken to, it was these activity clusters that absolute time was allocated to.

Qualitative

Richer contextual meaning on the other hand was provided within one week of the event through a semi-structured interview with the event manager. In the first instance, Event 1 and the management of its match day implementation phase was classified by the event manager as being 'typical' with 'no abnormal encounters' occurring. By comparison Event 2 was reported to be 'very quiet' with 'no phones ringing'. One explanation for this was that this game was played on a Saturday evening rather than the usual Friday evening. Furthermore, Event 2 was scheduled at the same time as the more popular professional football team playing in the city. This might have had implications for less media interest, a lower spectator attendance and, from the manager's perspective, may explain why Event 2 entailed a slightly shorter working day (by 75 minutes).

When the event manager was presented with the Event 1 chronological analysis of his activities, he thought that the findings provided him with an accurate picture of what he considered to be his normal event implementation activities. Developed over 5 years of basketball event management experience, match day was largely perceived as a time driven set of routines (chronos) that involved an integration of networked systems (kairos). These, for example, included established and refined standard operating practices relating to detailed media schedules, adherence to National League requirements (*e.g.*, precise location of advertising boards) as well as multi-agency approved emergency procedures. Given the manager's experience and the externally established requirements, he was not surprised to learn that less than 6% of his day was spent on *Problem Solving* tasks suggesting a closed system and relatively controlled semi-predictable and practiced working environment (chronos).

However, further analysis of the event manager's perceived level of reactivity, implied less control of his time usage than he realized. For example, more than 40% of activities in the 30 minutes before the match started were either rated as being high/medium reactive, namely initiated by some thing or someone else. In addition, it appeared that other stakeholder agendas, particularly the media and match officials, frequently dictated the urgency of tasks performed by the manager.

Post-hoc analysis provided numerous other incidents, both positive and negative, on which to appraise the event manager's time usage. Due to their unique and limited transferability, the specifics of these findings will not be presented here. Rather, more general processes and key lessons of time management analysis will now be discussed to further progress future studies in managing time.

The event manager was surprised to see that his time usage in Event 1 had revealed a high proportion of time spent on low importance and low urgency activities (*e.g.*, social conversation). In appraising such activities on an individual and contextual basis, the parameters of each task and loci of control were further explored and, where appropriate, a potential cause and solution proposed. As an example, retrospectively focussing on specific activities, the event manager explained that there were some tasks of varying importance that he felt he 'simply had to do or personally chose to do'. These included dealing with certain *Financial Activities* (regarded as confidential information as he was the only signatory for player payments and business deals) and the *Match Day Programme* (very strong personal marketing background and competence). Conversely, relating to other identified low important tasks he was surprised just how much time was 'not particularly productive for my position' (*e.g.*, time spent on low important, low urgency tasks such as *Watching the Game*) and how little *Delegation* he actually did. While he had the capability and motivation to carry out all tasks efficiently, the question was whether he was the only person who could carry them out? The answer in many cases was that, other people could be trained and the tasks delegated, although the number and type of staff available was a limiting factor (*i.e.*, a small full-time staff team and large array of volunteers). From Kerzner's (2001) analysis of reduced personal productivity, low important tasks that are not delegated may be classified as 'internal time robbers' (those time wasting factors that are directly under the manager's control). In this case, it was determined that there were a variety of tasks that the event manager need not necessarily do himself and for which he, in principle, possessed the locus of control to personally change, should he choose to.

Unbeknown to the investigators until the Event 2 post-event interview, the event manager had instigated some conscious actions in the close season to improve his personal productivity as a result of the earlier findings. First, he employed a part-time administrator to whom he delegated some low important reactive match day tasks and, secondly, he decided to spend less time courtside *Watching the Game* and use saved time to more productively develop strategic sponsorship partnerships. This is reflected in the differences identified between the Event 1 and 2 findings, namely a reduction of 9% of his total time on *Supervising/Watching the Game* and a new activity entitled *Proactive Networking*. This new activity, which included a post-event hospitality function, now accounted for 12% of his total work day and illustrates the very essence of kairos, where independent actors (the players, the event manager, potential/current sponsors and others) were proactively invited to an exclusive

post-game activity (interactive opportunity structure), thereby, increasing the probability of a mutually beneficial outcome (*i.e.*, future business relationships).

How productive these changes are must be appraised over a longer period of time. However, the important point is that the event manager's environment and the focus of his behaviour was more readily beginning to mirror the more holistic and dual approach suggested by Hedaa and Törnroos (2002).

Elaborating further and as highlighted in Figure 3, ideally the two triangles of chronos and kairos need to interact for memorable spectator entertainment to occur. On the one hand, the chronos triangle can be illustrated by pre-scripted announcements and routines prepared by capable and highly motivated autonomous actors (event manager, media personnel, basketball players, match announcer, cheerleaders etc). On the other hand, the successful implementation of these routines *in situ* is dependent upon comprehension of the kairos triangle. Potentially, these pre-event routines could be implemented on an *ad hoc* basis at opportunistic moments, and their appropriate synergy and coordination be left to good fortune and luck. Conversely, the event manager began to proactively determine and manage the opportunities when the routines were executed. This required a more holistic understanding of the environment and a more professional coordination of the required heteronomous actors performing, either in sequence or simultaneously. While some of these opportunities and time frames were predictable (*e.g.*, pre-game entertainment where cheerleader dances, music and lights can be rehearsed to precede player entrances), others need to react to appropriate triggers. Thus, additional basketball entertainment could be created to accompany game incidents such as immediately an opponent is sent off, scoring of a three point play, time

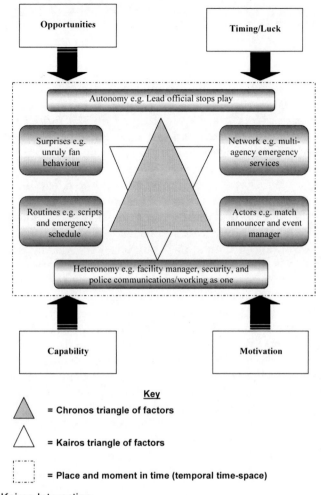

Fig. 3. Chonos and Kairos Interaction
Source: Adapted from Hedaa, and Törnroos, 2002

outs or even once a particular player mile-stone was reached. Many of these incidents are likely to take place across a season or game, but exactly when is unknown. This is where the ability to react, the very essence of kairology and appropriate timing, becomes essential in the event management environment. This entails determining what the likely surprises are and making sure that the performers are readily available so as to respond to predetermined moments or actions. This should not just occur to create positive outcomes, but also to react

to negative crises as seen with the development of incident reporting processes.

In exploring other contextual time usage patterns of the case study, it was determined that one of the key causes and triggers to the time wasting activities were the frequent 'interruptions by others'. While these are often inevitable and unpredictable under match day conditions the frequency, likelihood and duration of them can be reduced. According to Kerzner (2001), this represents an external time wasting factor, where the manager can control the proximity and synchronous nature of such interactions (opportunity structure). At one extreme, he can consciously make himself accessible to all stakeholders (*e.g.*, placing himself on courtside), or at the opposite extreme make himself unavailable unless an absolute emergency occurs (*e.g.*, closing his office door).

In practice, further analysis of external interruptions revealed that there were two types – in person and through technology. Although the findings reveal that the event manager limited the duration of these interruptions, this only happened once they occurred. For example, externally generated face-to-face and technology communications occupied relatively short durations of less than 10 minutes. In the case of telephone usage, the highest average individual call (incoming or outgoing) was just 2 minutes 30 seconds (Event 1), yet cumulatively, they occupied 8% of the total time spent at each event. Focussing more specifically on computer usage, which was most heavily used in the pre-match phase, the control of technology interruptions was less than ideal. This finding seems to support the work of Wilson (2005), who suggests a societal 'adult problem of infomania', where 62% of adult workers appear to be addicted to checking incoming technology messages as soon as they arrive, with 50% answering them immediately. Regardless of whether this 'always on' technology included mobile phones or computers, Odih and Knights (2002, p. 74), similarly endorse the point that

> We allow ourselves to be controlled by its routines, feeling the ever demanding pressures to provide instant responses to messages.

Indeed, the Event manager was aware of the time wasting element of emails, in general, reporting that he typically 'received anything from 50 to 80 emails per day, of which 45 were often junk!' However, he was not aware of an apparent state of permanent readiness to react to so many types of technology (information clutter) at any moment in time.

From this analysis a personal development action plan was developed in consultation with the event manager. This covered everything from improved self-discipline (simply turning off the mobile at particular times), streamlining the number of communication media used, educating others as to what should and should not be communicated by whom and how, establishing protected time for important tasks, developing screening mechanisms and working predominantly to his agenda rather than in response to instantaneous demands and expectations of others.

Regarding the very long hours experienced on match days, the event manager expressed that this was the 'expected nature of the business' and given that he was also the CEO, it was in effect both his livelihood and passion in life. He recognized that there was very little distinction between his work and personal life and by adopting wireless technology communications demonstrated further adherence to both monochronic and polychronic work patterns (Lee and Liebenau, 2002). However, when asked the questions of what would happen if he was ill, and could anyone else within his organization do the job, he realized that continuous commitment to these long hours was perhaps not rational in a business sense, but one taken through personal choice and enrichment. This highlights the importance of understanding cultural and personal needs that must drive the method of data collection, appraisal of time usage, as well as ownership of the ensuing action plan.

RECOMMENDED DATA COLLECTION AND ANALYSIS

Collection

In discussion with the manager, the preferred method of independent analysis of his match day use of time was for the researcher to accompany him to all activities. Initial concern was expressed about the researcher's shadowing presence, and its potential to influence the behaviours of both the event manager and co-worker and hence validity of data collection. On the one hand, the event manager elaborated that this was not an alien experience to him or his small group of working colleagues, as numerous placement students had shadowed his every move on match day over the last two years. On the other hand, while a video would be the ideal recorder of information, particularly for *post-hoc* analysis purposes, it was agreed that this might be too intrusive and given the uniqueness 'some staff might play up to the camera'. To this end and to try and create as normal as possible work environment it was decided that the primary method of data collection was for the researcher to record all information via pencil, paper and stopwatch.

The actual method and tabulation of data recording used and recommended was via a work sheet as illustrated in Table 2.

So as not to overcomplicate live data collection, only a few elements were recorded *in situ*, namely

- a concise description of the activity (*e.g.*, writing cheques) as well as identifying any external actor involved (*e.g.*, EM calls A – abbreviations used to respect personal identities)
- the start and finish time of the activity (to the nearest minute)
- the place of the activity undertaken (*e.g.*, office, courtside, entrance foyer).

Other information, such as the activity duration and Category Classification were added afterwards, and context appraisal information such as the level of reactivity, importance, urgency and potential to delegate were inserted into the relevant columns by the event manager before the *post-hoc* interview. A Dictaphone was also used to back up the live data.

For research, this method of *in situ* data collection proved to be a very user-friendly and 'comprehensive way of recording actual practice' (researcher opinion). From the perspective of a researcher, it was stated as being 'extremely easy to identify specific activities and accurately record data' and involved 'barely any time to administer'. To illustrate, Event 1 (11 hours 30 minutes) was recorded using just nine A4 pages. Importantly, the information was very easy to understand directly after the event for both the researcher and event manager. Consequently, both the researcher and the manager felt that the focus on individual tasks with its sequential written presentation permitted 'contextual memory and understanding to which further personal interpretations could be added'. Where clarification was required, reference to the Dictaphone recording was possible.

Regarding the validity of the time usage parameters, as illustrated in Table 1, there were few differences between the Event 1 and Event 2 cluster and individual percentage data summaries. Where individual differences did exist, *i.e.*, in *Proactive Networking* (+12% change) and *Supervising/Watching the Game* (−9%), these could be explained by learning effects and conscientious actions taken to improve personal productivity. On this basis and the fact that chronological time was measured via absolute clock time, it is asserted that this method of data collection appears to be an accurate and reliable method of time oriented data collection.

Analysis

Once the *in situ* data had been recorded, the researcher was able to calculate the duration of each activity and categorize its nature. While the former was very easy to achieve, unfortunately the latter was not. As no previous study of category analysis had been found, a subjective classification of clusters was formed. As identified in the quantitative

Table 2 Worksheet example of chronological data recording (event 1)

Time	Duration	Place	Nature of activity	Category Classification	Reactive	Importance	Urgency	Potential to Delegate	Other Comments
10.15–10.20		Office	EM calls trying to contact A about tonight game – left message		L / M / H	L / M / H	L / M / H	L / M / H	
10.20–10.25		Office	Reading emails in front of computer		L / M / H	L / M / H	L / M / H	L / M / H	
10.25–10.29		Office	Writing cheques – today is wages day for players		L / M / H	L / M / H	L / M / H	L / M / H	
10.29–10.33		Office	Reading the "things to do" list; analysing and deciding what and when to do		L / M / H	L / M / H	L / M / H	L / M / H	
10.33–10.36		Office	B comes into office. Talks about season tickets		L / M / H	L / M / H	L / M / H	L / M / H	
10.36–10.40		Office	Writing cheques and makes payments to creditors.		L / M / H	L / M / H	L / M / H	L / M / H	
10.40–10.43		Office	1st call received on mobile phone – personal, business		L / M / H	L / M / H	L / M / H	L / M / H	
10.43–10.45		Office	Writing cheques		L / M / H	L / M / H	L / M / H	L / M / H	
10.45–10.50		Office	B comes into office. Talks about marketing and season tickets (home game returns)		L / M / H	L / M / H	L / M / H	L / M / H	

Key:
L = Low; M = Medium; H = High
N.B. Event Manager (EM) circles most appropriate L / M / H per cell

findings, the hierarchical cluster categorization used in this particular research was based upon identified 'event specific tasks' (highest order), 'management functions/ activities', 'communications' and 'other activities' (lowest order). Clearly this hierarchical categorization has been driven by the findings of the case study and, in particular, the activities encountered in Event 1.

This provides the potential for inconsistent categorization and potentially overlapping categories. While the boundary between categorizations could always be subjective, in the current scenario the researchers always resorted to the higher order category, where overlaps or multi-tasking occurred.

It is recommended that further research of sport events be undertaken to establish a more robust and consistent categorisation framework. The strength of the current methodology and findings are that they serve to describe the exact nature of activity as well as location/actors and can be re-categorised in the future if necessary.

As a technique, performance of the post-event interview within one week of the event, received strong support from both the researcher and manager. The event manager stated that 'anything longer may lead to me forgetting the exact context, and I certainly wouldn't want to do it directly after the event due to fatigue!'

Conversely, the subjective nature of importance and reactivity interpretation could have made the categorization process difficult. For example, what is the term length and unit of importance (*e.g.*, short term may relate as important to the event on the night, whereas long term could be considered at the business level and over 5 years)? How might importance be measured, and from whose perspective? Similar anomalies could also have been encountered with the terms reactivity and urgency (*e.g.*, urgent from whose perspective?). However, in practice no such problems were experienced, instead the

manager fully supported the system proposed and circled the most appropriate response on the work sheet on a task-by-task basis. This task was reported to take less than 30 minutes for each event.

Once categorized, both the researcher and event manager suggested that it was a case of contextually identifying the locus of control into either external (*e.g.*, interruptions, bureaucracy, incomplete information, poor organization by others) or internal time wasting factors (*e.g.*, procrastination, inability to say no or to delegate, personal organization, lack of planning), and then developing a personal action plan to address the likely problem cause (*e.g.*, establish tactics to influence others or methods to improve personal planning).

Overall the event manager reported that this time analysis had been enlightening both at the personal and organizational level. In his words 'It is rare that self-monitoring of time usage is ever undertaken and even rarer that this is done independently'. Short term management-initiated changes that were implemented as a result of these particular findings included

- A part-time administrative appointment for match day. This position was used to screen many of the reactive and peripheral time wasting tasks such as fan requests.
- Personal changes in how the event manager used his time and technology in general.
- Organization of time management training for all staff, where systems were established to appraise, monitor and evaluate individual time usage on a six monthly basis.

CONCLUSION

In looking to the future, Purser (2002) cites Helprin (1999, pp. 264–265), who suggested that with technological developments

The man of 2016 ... [will] no longer [be] separated from anyone. Any of his acquaintances may step into his study at will – possibly twenty, thirty, forty, or fifty a day. If not constantly interrupted, he is at least continually subject to interruption, and thus the threshold of what is urgent drops commensurately.... No matter how petty a matter, a co-worker can appear to the man of 2016 in a trice. Screening devices or not, the modern paradigm is one of time filled to the brim. Potential has always been the overload of will, and the man of the first paradigm finds himself distracted and drawn in different directions a hundred times a day.

This exploratory case study has begun the process of trying to better understand and improve the time management practices in the sports event industry. While previous work on time management was unduly focussed on the quantitative and chronological nature of time (Adam, Whipp and Sabelis, 2002), this research determined a more contextual and temporal understanding of time, applying Hedaa and Törnroos's (2002) chronos and kairos approaches to a single case study.

More specifically this case study provided an in-depth focus on the actual practices of an experienced event manager at two National Basketball League events. Its' findings included

- Identification of which activities the event manager normally engages with on the implementation day of a match,
- Determination and evaluation of the context as well as actual time usage of an event manager,
- The importance of controlling interruptions particularly technology in a world of instantaneous communications,
- Introduction of a user-friendly and non-invasive method of data collection and analysis that could be applied to other event settings.

From these findings, it is recommended that as good management practice regular independent analysis of time usage and accompanied context are undertaken at both the individual and the organizational level. From such analysis, poor and excellent practice can be identified, and through consultation and facilitation, personal progressive management plans can be designed to use time more productively.

From a research perspective this is just the start. This research is clearly limited by its focus on just one event type and on one manager. Further research is required across a broad spectrum of event typologies, so that a more acceptable, robust and consistent categorization framework of time can be developed. Greater focus is additionally recommended towards the relatively new approach of kairos and the proactive nature of timing to increase as well as decrease the likelihood of specific outcomes occurring. For example, what are the trigger events of a crisis or a truly memorable atmosphere and how might they be recognized, controlled and enhanced?

Time is a scarce resource and with the event industry renowned for very long work hours and high degrees of burnout (Goldblatt, 2005) it is indeed time to manage time.

REFERENCES

Adam, B., Whipp, R. and Sabelis, I. (2002) Choreographing time and management: traditions, developments, and opportunities. In: R. Whipp, B. Adam and I. Sabelis (eds) *Making Time and Management in Modern Organizations*, New York, Oxford University Press, pp. 1–28.

Blake, P. (2004) *Unpublished Informal Interview with British Basketball League Chairman Director*, September 10 2004.

Bowdin, G., McDonnell, I., Allen, J. and O'Toole, W. (2003) *Events Management*, Oxford, Elsevier Butterworth-Heinemann.

Bunce, D. (2000) *Sport & Leisure Facility Management – Distance Learning Pack*. MSc Sport Management, Newcastle, Northumbria University.

Cole,G. A. (1993) *Management Theory and Practice*, fourth edition, London, DP Publications.

Elliott, D. and Smith, D. (1993) Football Stadia Disasters in the United Kingdom: Learning from Tragedy? *Industrial & Environmental Crisis Quarterly*, **7**(3), 205–229.

Emery, P. (1997) *The Management of Major Sport Events*. Unpublished Master of Business Administration Thesis, Durham University.

Emery, P. (2003) Sport event management. In: L. Trenberth(ed) *Managing the Business of Sport*, Palmerston North, Dunmore Press, pp. 269 – 292.

Getz, D. (1997) *Event Management & Event Tourism*, New York, Cognizant Communication Corporation.

Godefroy, C. H. and Clark, J. (1990) *The Complete Time Management System*, London, Piatkus Publishers.

Goldblatt, J. J. (2005) *Special Events ~ Event Leadership for a New World*, fourth edition, New Jersey, John Wiley & Sons.

Hartley, H. J. (1997) Hillsborough – A Disaster Unfolds: Implications for Managing Mass Sport and Recreation Events. *Sport and The Law Journal*, **5**(1).

Harvard Business School Press (2005) *Time Management: Increase your Personal Productivity and Effectiveness*, Boston, Harvard Business School Press.

Kerzner, H. (2001) *Project Management: A Systems Approach to Planning, Scheduling, and Controlling*, seventh edition, New York, John Wiley & Sons.

Hedaa, L. and Törnroos, J. (2002) Towards a Theory of Timing: Kairology in Business Networks, in: R. Whipp, B. Adam and I. Sabelis (ed) *Making Time and Management in Modern Organizations*, New York, Oxford University Press, pp. 31–45.

Lee, H. and Liebenau, J. (2002) A new time discipline: managing virtual work environments, in: R. Whipp, B. Adam and I. Sabelis (eds) *Making Time and Management in Modern Organizations*, New York, Oxford University Press, pp. 126–144.

Lussier, R. N. and Kimball, D. (2004) *Sport Management Principles, Applications, Skill Development*, Mason, Ohio, Thomson/South Western.

Murray, M. (1995) How Atlanta Plans to Win Gold. *Event Organiser*, **23**, pp. 16–17.

O'Brien, S. (2006) Try Slowing Down to Survive. *Herald Sun Newspaper*, Feb 24 2006, p. 20.

Odih, P. and Knights, D. (2002) Now's the time! Consumption and time-space disruptions in post-modern virtual worlds, in R. Whipp, B. Adam and I. Sabelis (eds) *Making Time and Management in Modern Organizations*, New York, Oxford University Press, pp. 61–76.

Peters, T. J. and Austin, N. K. (1985) *A Passion for Excellence: the Leadership Difference*, New York. Warner Books Inc.

Phillips, S. (2002) *Time Management*, Maidenhead, McGraw Hill.

Phillips, S. (2005) Games Holiday Order. *Herald Sun Newspaper*, 28 Oct 2005, p. 12

Purser, R. E. (2002) Contested presents: critical perspectives on 'real-time' management, in: R. Whipp, B. Adam and I. Sabelis (eds) *Making Time and Management in Modern Organizations*, New York, Oxford University Press, pp. 155–167.

Radu, A. (2005) *How Do Event Managers Utilise their time in the Planning, Implementation and Post-Event Phases of National 1st League Basketball Events*. Unpublished Masters in Sport Management Thesis, Northumbria University.

Silvers (2006) *Speaking of Events – Event Management as a Profession*, available at: http://www.juliasilvers.com/embok.htm

Torkildsen, G. (1999) *Leisure & Recreation Management*, fourth edition, London, E & F Spon.

Watt, D. C. (1994) *Leisure and Tourism Events Management & Organisation Manual*, Harlow, Longman.

Westerbeek, H., Smith, A., Turner, P., Emery, P., Green, C. and Van Leeuwen, L. (2005) *Managing Sport Facilities & Major Events*, Crows Nest, Allen & Unwinn.

Wilson, G. (2005) Infomaniacs Just Hi-Tech Dopes. *Herald Sun Newspaper*, April 23 2005, p. 2

The dynamics of successful events – the experts' perspective

Dr. John Ensor, Martin Robertson and Jane Ali-Knight

INTRODUCTION: FESTIVAL MANAGEMENT AND LEADERSHIP

In 1988, Getz and Frisby stressed the role of communication and understanding in the effectiveness of organizers' involvement with community run festivals. Getz (2000a) further stated that management understanding is vital to the success of events, and, later, that misunderstanding is central to their failure (2000b). Gursoy, Kim & Usyal (2004) reflected that the growth of festivals and events has encouraged a rise in the professionalism of their management, inclusive of a rising awareness of the social, environmental and economic impacts associated with these actions. Moreover the growth in festivals, which are so dependent on stakeholder relationships, has led to researchers to focus more on inter-organizational relationships and network structures (Stokes, 2006).

Recent years have also seen a rise in strategic planning and management of events in response to the adoption of a corporate culture by not-for-profit festivals (Spiropoulos, Gargalianos and Sotiriadou, 2006). McDonnell, Allen and O'Toole (1999) state that it is vital that managers identify stakeholders as part of a strategic planning process due to the various agendas a festival has to meet. Moreover, they state that festival organizers should consult with all stakeholders during the shaping of the vision, mission and goals of the event. As they highlight,

'It is no longer sufficient for an event to meet just the needs of its audience. It must also embrace a plethora of other requirements including government objectives and regulations, media requirements, sponsors' needs and community expectations' (p. 39).

The authors conclude that festivals and events are judged by their

'success in balancing the competing needs and interests of a diverse range of stakeholders' (p. 39).

Certainly it is generally recognized that effective strategic management of inter-organizational linkages are integral to the survival of any organization and that they can particularly enhance stability during periods of increased competition (Merrilees, Getz and O'Brien, 2005).

FESTIVALS IN EDINBURGH – A BACKGROUND

Edinburgh is host to fifteen diverse national and international festivals annually, as well as several community and participative events. These range from the prominent and internationally known Hogmanay and the Edinburgh International Festival (EIF) to lesser-known but equally important festivals such as The Beltane Fire Festival and the Scottish International Storytelling Festival. Covering two hundred and fifty one days of the year, festivals are clearly a vital part of Edinburgh's life. Principal impacts lie in the areas of cultural, social and economic benefits and civic pride.

A study commissioned by the key stakeholders in the cities' Festival framework (The City of Edinburgh Council; Scottish Enterprise Edinburgh and Lothians, Visit Scotland and EventScotland) in 2004 measured the economic impact of the festivals. The study aimed to update a previous report (Scottish Tourist Board, 1990) and confirm the collective success of the festivals in generating revenue for Edinburgh and Scotland. Results showed that from August 2004 to July 2005 Edinburgh's festivals attracted over 3.1 m attendances, and an estimated 1.4 m trips to the city. In Edinburgh and Scotland, they generated output of just under £170 m and £184 m,

respectively – £40 m in new income and support for 3200 full time equivalent jobs would not exist if the festivals did not take place. In Scotland as a whole, direct new income measured £51 m, resulting in 3900 full time equivalent jobs (SQW *et al.*, 2005). The multiplier effect on tourism businesses in the city is also significant with hotel occupancy rates typically soaring to 80–90% in the capital during the festival period and a net increase of £22.5 m for bars and restaurants.

FESTIVAL DIRECTORS IN EDINBURGH

In Edinburgh, the Festival Director is a key stakeholder and is often credited with having the strategic vision and stamina to grow and develop the festival, both in respect of its image and in the development of its artistic achievement. As example, in the case of the Edinburgh International Festival, Sir Brian McMaster stepped down, in September 2006, after fifteen years as Festival Director of the Edinburgh International Festival (EIF) and has moved to be on the board of the Manchester International Festival. Commenting on this move Catherine Lockerbie, Director of the Edinburgh International Book Festival, reinforces the importance and collateral value of festival leaders when she says

I think it is enormously clever of them. I would have tried to do the same in their position. There's nobody better than Brian to advise them. It's cheeky and rational, and an extremely smart move (Cornwell, 2006).

Jonathan Mills has recently taken up the position at EIF and will be responsible for planning the Festival this year (2007) and onwards. One of Australia's most experienced festival directors, his previous posts have included Artistic Director of the Melbourne International Arts Festival, the Melbourne Federation Festival, the Melbourne Millennium Eve celebrations and the Brisbane Biennial

International Music Festival. On taking up the post he expounded the cultural legacy of his role saying,

> "Edinburgh International Festival is a great Festival in a gracious city. From the moment it was founded EIF has been a source of inspiration for similar celebrations of the performing arts throughout the world. I am both excited and humbled to be offered the opportunity to build on the twin traditions of excellence and innovation established by this wonderful event over the last 60 years." (www.eif.co.uk).

More recently Paul Gudgin, Director of the Edinburgh Festival Fringe since 1999, has received a great deal of national news coverage with his announcement to step down from his role at the start of the 2007 event. The Edinburgh Fringe Festival announced that in 2006 the festival featured 28,014 performances within 1,867 shows in 261 venues (www.edfring.com). It is perhaps no surprise then that the arrival of the new Director, Jon Morgan, is keenly anticipated and reported within UK media.

EDINBURGH CULTURAL POLICY DIRECTION AND FESTIVALS

The Edinburgh Festivals Strategy was launched in 2001 – it followed the Cultural Policy (1999) for the City and was commissioned and funded by the City of Edinburgh Council with financial support from the Scottish Arts Council and Scottish Enterprise Edinburgh and Lothian. It was developed in tandem with an Events Strategy for the city, which was launched in December 2002. Both strategies are evidence of Edinburgh's commitment to position itself as the 'Festival City' and their realization of the importance of festival and events to the cultural and economic viability of Edinburgh.

The Festivals Strategy was a result of personal interviews with key stakeholders in the Festivals and Tourism industry,

discussion with core groups (*i.e.*, Joint Festivals Working Group) and extensive desk research, which included benchmarking the Edinburgh festivals against other cities. Importantly, the festival directors had direct input to this process throughout and the strategy grew out of their needs for a more co-ordinated and tactical approach to festival provision in the city.

The strategy then recognized the need for a shared vision, which the City of Edinburgh Council (CEC), the various festivals and other interested parties could sign up with a common plan of action. The festival directors were integral both to its development and its final conclusion, the Festivals Strategy Action Plan. Paul Gudgin, Director of the Edinburgh Festival Fringe during these important years of change, commented that the action plan helped 'to foster closer working relationships across many of their departments, and between all the festivals and a number of other key agencies.' (Edinburgh Festival Fringe, Annual Report 2002).

The key aims of the strategy included delivery of festivals and events throughout the year as well as maintenance of the quality and diversity of the summer festival programme for which Edinburgh is best known; ensuring viability in respect of the social, commercial, artistic and cultural needs and objectives of the city; linking learning and cultural development around festival provision; ensuring a strong relationship between the City of Edinburgh Council and the festival providers; effective financial and marketing support and marketing relationship between local and national public cultural and tourism agencies and 'interconnectedness between the festivals, enabling co-operation, joint initiatives and the sharing of resources, stimulating a positive sense of creative competition'.

In 2005, and following the strategy document, a report was commissioned to examine the competitiveness of Edinburgh's eleven principle Festivals. Published in 2006, the report compared Edinburgh's festivals with that of twenty festivals in other cities around the world. Instigated by the same stakeholders as the Festival Strategy, and with additional funding from the Scottish Executive, the report – aptly named 'Thundering Hooves' – was a response to an increasingly visible competition from both within UK and from overseas, and the 'increasing use of cultural programming (*i.e.*, festivals and events) as strategic devices to promote tourism and to build brand-identity of the cities or regions where they are located' (AEA Consulting, 2006). This last factor is evidenced most particularly within the United Kingdom by the engagement of cities, such as Liverpool and Manchester, in post-industrial regeneration activities, wherein a key element in their competitiveness is the extensive level of public and private investment placed in creating both the festival and events themselves and the infrastructure and development they require. Liverpool's success in achieving City of Culture status for 2008 was very much seen as emulation of the urban regeneration and cultural revitalizing activities of Glasgow following its hosting of the award. Similarly, in Manchester, festival strategy activity has been spearheaded by the Manchester International Festival 2007, which claims to be 'the world's first international festival of original, new work' at a cost of £6 million (www.manchesterinternationalfestival.com), which is a considerable outlay for an event with, as yet, no history.

In its conclusion the Thundering Hooves' report reflected that, while Edinburgh benefits from a world class festival presence with an incomparable breadth and extent of festival activity, it was also vulnerable to competition from other city's with festivals. It put forward fourteen main points of action in response to this. Significantly, it also highlighted the advantage that Edinburgh had over other city's hosting festivals. Core to this advantage was the co-operation that

existed between the festival directors and the local and national public/private sector agencies. Perhaps for this reason, there may be a clear distinction in the relationship that exists between the festival director in Edinburgh and their working environment with that of festival directors and their work elsewhere in the UK.

Thundering Hooves (2006) has left a strong legacy for festivals in Edinburgh. In following through the key recommendations of the report, as well as lobbying the government and the private sector for much needed increases in funding, it has also heralded the graduation of the new organization 'Festivals Edinburgh' as advocates for the festivals,

METHODOLOGY

Repertory grids were employed to allow the identification of key constructs that the sample of festival leaders found within their own professional environment. Repertory grid technique is an established psychological tool that has been used for over 40 years and has an extensive range of business application (Stewart and Stewart, 1981; Easterby–Smith *et al.*, 1996). The Repertory Grid technique has its foundations in the Personal Construct Theory (PCT) developed by Kelly (1955) and is described as serving to discover an interviewees' construct by way of conversation (Fransella and Bannister, 1977; Jankowicz, 1995; Jankowicz, 2003; Pike, 2003; Selby, 2004). While originally developed for application to single respondents, it has great flexibility and is not necessarily limited by any particular sample size (Pike, 2003). By involving more than one person the process serves to provide recurring constructs of knowledge based on consensus (Selby, 2004).

Within the field of tourism research, this process has been applied for evaluation of destination image in the minds of potential and immediate consumers Coshal (2000). Pike (2003) concluded that the techniques are likely to offer constructs valuable to the decision making process within an organization (or destination).

In the analysis of festival and arts provision, the use of repertory grid analysis is less common. Where it is applied, it is more commonly associated with ascertaining market knowledge. Jansen–Verbeke and Rekom (1996) used repertory grid analysis to identify the motivation of visitors to museums (in Rotterdam) for the purpose of more effective marketing of the urban tourism product. They conclude that the motivation construct for visiting museums is convergent with the needs of visiting a city, and thus not distinct from it. Caldwell and Coshall (2002) used repertory grid analysis to measure brand association of museums and galleries. They found that there were a relatively small number of constructs, concluding that efforts by museums and art galleries to create brand associations have not been effective with customers failing 'to differentiate between them on the basis of anything but functional benefits' (390).

Canning and Home (2006) used repertory grid as a consultation devise with selected community groups in Sheffield to evaluate the best form of museum and art gallery provision. Importantly, their work highlights both the robust nature of repertory grid data and the pragmatic value to both community and government policy makers in offering meaningful data, which bridges the governments desire for quantitative data, while dealing with issues suited to qualitative methodologies.

APPROACH TO THE REPERTORY GRID INTERVIEWS

In this research study the interviewees were asked, prior to the interview, to identify five festivals within the UK with which they were acquainted. These were used as the elements of the repertory grid. Each of the interviewees, therefore, had a specific and unique set of elements. In generating constructs the 'triading' method was used (Coshall, 2000; Pike, 2003; Selby, 2004). Each interview took at least an hour. A follow up session then took place, where each respondent rated the constructs on the repertory grid that the researcher had drawn up as a result of the initial interview.

Honey's (1979) method of content analysis was used in preference to a more highly statistical approach. This has the advantage in allowing different constructs across a sample to be combined, while at the same time allowing the researcher to make use of individual meanings expressed through the rating given on any single repertory grid.

Jankowicz characterizes this form of content analysis as being a technique that,

'assumes that what we're interested in is each individual's personal understanding of the topic in question, and treats each construct offered by the individual as more closely related, or less closely related to the overall issues he has in mind when thinking about the topic' (2003, p. 170).

Although respondents chose different elements (festivals), they were asked to rate how creative/uncreative each event was from their perspective. All the other constructs they rated on their repertory grid could then be analysed in terms of how close or removed it was from this, as Jankowicz (2003) calls it, 'overall summary' construct. In order to do this, each construct is labelled with two indices that have been calculated. The first index is the percentage similarity score. A score of 100% indicates that the rating on that particular construct match exactly the ratings on the overall summary construct. A rating of 50% would indicate that the ratings were substantially different. Recognizing that similarity scores are relative Honey's procedure then requires the researcher to note whether the similarity score on a construct is in the individual's highest scoring third of constructs, the intermediate third or the lowest third. Thus High, Intermediate or Low (H–I–L) values are allocated to each construct accordingly as the second index. Accordingly, each construct by this stage then has a percentage similarity score (SIM) and an H–I–L value as well as its own reference code.

Constructs are then allocated to categories as in a generic approach to content analysis. Each construct is compared with the other. Constructs that are the same in some way are placed together under a single category. Constructs that are different from existing categories are placed separately and start the formation of a new category. This process continues until all constructs have been classified.

The Honey (1979) procedure allows constructs to be placed within a category in order with those closest to the summary construct at the top of the list. The H–I–L scores also allow the researcher to establish exactly how important individuals rated the constructs within a particular category. In carrying out this procedure the percentage similarity score has to be calculated for the construct in both its positive and reversed relationship with the summary construct. The percentage similarity score that is closest, whether positive or reversed is the score that is used when attaching the index score and the H–I–L value. This methodology was appropriate to the needs of the research.

Findings

The findings from the research are summarized in Table 1. Each category is shown in the order of rank. Construct categories that were ambiguous or contradictory have not been included in the findings. The percentage total for each construct category (from all constructs) is shown in the fourth column. This adds to 47% of all constructs. Each of these construct is then discussed and represented in Figures 1 to 6 below.

The construct category is in the centre of each figure and the subcategories that were elicited from this (where they exist) are its extensions from this. The constructs that formed the subcategories are shown as

Table 1 List of Key Categories and Sub Categories Arising out of Honey's Content Analysis of the Total Sample's Repertory Grids

Ranking	Categories	Sub-categories	% of Total Constructs	% of Constructs rated High on SIM score
1.	Leadership	• Independence • Freedom • Culture	18	42
2.	Focus	• Audience driven • Artistically driven	14	43
3.	Relationship with local community	• Mutual appreciation • Lack of major divisions within local community	10	50
4.	Decision making style	• Decision making style • Influences on decisions	6	67
5.	Funding		5	60
6.	History of the festival		4	50

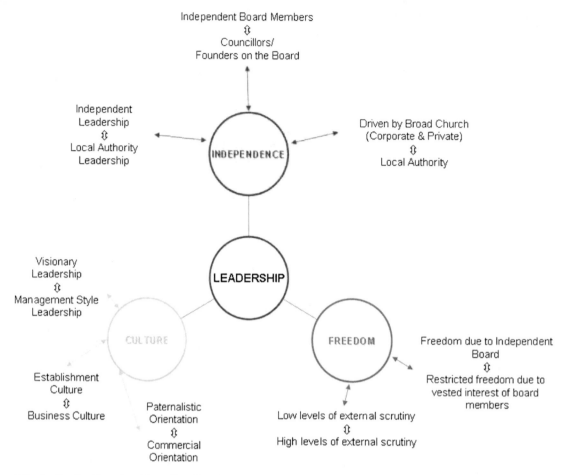

Fig. 1. Festival Leadership – Key Constructs

branches from these extensions. An extended discussion is made for the first category, Leadership, to animate interpretation of the diagrammatical representation of the construct categories.

Leadership

18% of all the constructs identified through the repertory grid interviews fell into this category. Of those constructs, 38% rated high on the individual's similarity score with a further 16% rated as intermediate. Three key aspects of leadership were identified through the repertory grid. There were independence leadership, freedom from external pressure and a 'non-business' orientated culture. The key constructs that underpin this category are indicated in Figure 1.

Each key aspect of the construct category *Leadership, i.e.,* independence, freedom and culture, had clear subcategories. A perception of involvement from outside the festival organizations or expectations from outside the festival leader's role was palpable in each of these. *Independence* was perceived as being effected by the form and extent of shared decision making, particularly where it involved local authority organizations. The perception of *Freedom* in respect of leadership was again related to the influence of external bodies, although not necessarily ones from local authority agencies. The *Culture* of the festival environment was viewed as influencing leadership in respect of three key constructs. The first of these was the opposing (if necessary and pragmatic) balance between visionary and more structured management styles. The second indicates perception of an established (historic) culture as distinct from (a perhaps emergent) business culture as effecting the leadership. In the third, paternalistic orientation was viewed in respect of its relation to commercial origin.

Focus

14% of all the constructs identified through the repertory grid interviews fell into this category. Of those constructs, 43% rated high on the individual's similarity score with a further 29% rated as intermediate. The construct identified two separate but not necessarily mutual exclusive foci that successful festivals pursue. These are to be audience driven and/or artistically driven. The key constructs that underpin this category are indicated in Figure 2.

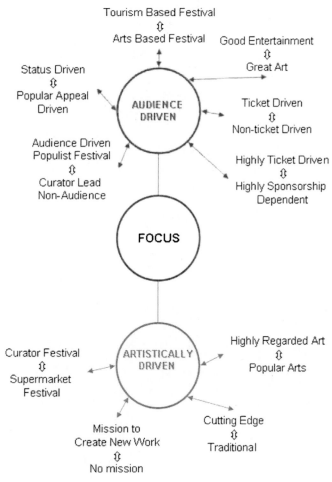

Fig. 2. Festival Focus – Key Constructs

Relationship with Local Community

10 % of all the constructs identified through the repertory grid interviews fell into this category. Of those constructs, 50% rated high on the individual's similarity score with a further 10% rated as intermediate. Two aspects emerged from the analysis. These are indicated in Figure 3.

Decision-making

6% of all the constructs identified through the repertory grid interviews fell into this category. Of those constructs, 67% rated high on the individual's similarity score with a further 33% rated as intermediate. The constructs represent two aspects of decision-making. First, the decision-making

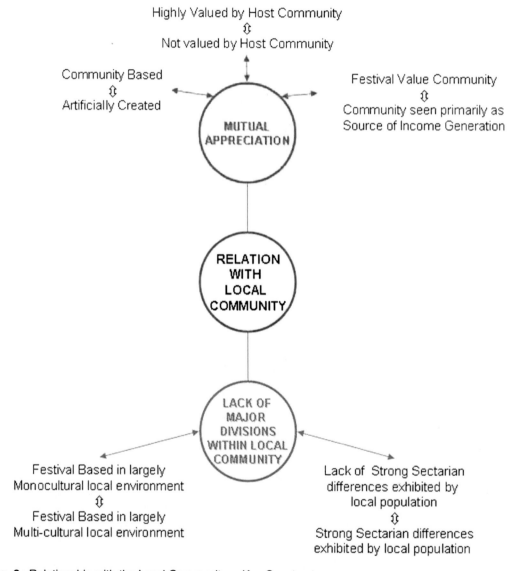

Fig. 3. Relationship with the Local Community – Key Constructs

style exhibited by successful festivals and secondly the key sources of influence on decision making. The key constructs that underpin this category are indicated in Figure 4.

Funding

5% of all the constructs identified through the repertory grid interviews fell into this category. Of those constructs, 60% rated high on the individual's similarity score with a further 40% rated as intermediate. The majority of responses suggest that festivals are more likely to be successful if they are supported with public money. The key constructs that support this view are indicated in Figure 5.

HISTORY OF THE FESTIVAL

4% of all the constructs identified through the repertory grid interviews fell into this category. Of those constructs, 75% rated high on the individual's similarity score with a further 25% rated as intermediate. Successful festivals were associated with them having heritage and tradition. While this is an interesting perspective, especially when linked with second-generation leadership, it may also just be reporting the obvious. Festivals that have longevity by default have demonstrated a key aspect of success. It may not, however, be an indicator of future success. The key constructs identified in this category are indicated in Figure 6.

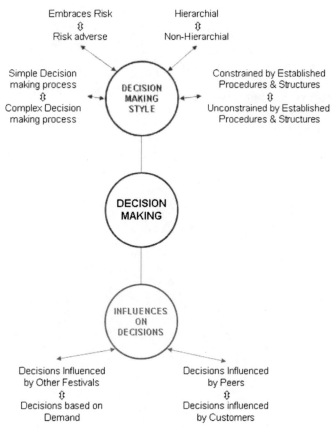

Fig. 4. Decision-Making – Key Constructs

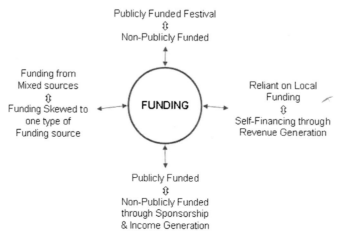

Fig. 5. Funding – Key Constructs

DISCUSSION AND CONCLUSIONS

Six key categories emerged from the constructs identified in this research. These are leadership, focus, relationship with the community, decision-making, funding and history of the festival. The constructs formed have in many ways confirmed what would be expected from the perception of those leading festivals in a city which, as the Thundering Hooves (2006) report indicated, have unprecedented level of stakeholder involvement in their provision.

However, the key sub-constructs may offer further insight as to what affects the creative and innovative potential of these festivals. At a time when, as the report indicates, competition between cities as host for innovative and attractive festivals is increasing markedly, this is a potentially valuable tool.

This investigatory analysis indicates that the leaders themselves consider that the leadership constructs of independence, freedom and culture of the organizational environment are the most significant characteristics effecting the creativity and

Fig. 6. History – Key Constructs

innovativeness of a festival. The second and third highest rated construct categories, Festival Focus (Fig. 2) and Relationship with Community (Fig. 3), respectively, can be seen to indicate a highly developed perception by festival directors of the issues affecting support in ensuring creativity and innovation for a festival; while Decision-Making Style (Fig. 4), Funding (Fig. 5) and History of the Festival (Fig. 6) may be seen as the perception of what it is that effects capacity to affect change. The constructs of the subcategories further highlight the challenges the respondents perceive.

Within the world of festival management 'successful managers of creativity are the brokers who add value to the creative process by directing the traffic of ideas and resources, and by "matching" ideas, individuals and organizational tasks' (Bilton and Leary, 2002: 62). In Edinburgh, the development history of the festivals per se, their accompanying strategies and the growth of organizational networks do indicate that creative brokering (Bilton and Leary, 2002) is at their heart. Similarly, the constructs elicited in the research undertaken here certainly indicate that festival leaders of the large events involved see these elements as vital. The constructs also indicate the factors that leaders perceive as constricting as well as alleviating the progress of this creativity. Given the pronouncements of the Thundering Hooves report, competitiveness requires proficiency in creative brokering.

The repertory grid technique is a valuable tool for evaluating perceptions of a particular target group. As Canning and Holmes (2006: 295) state, the application of this technique is particularly suited to groups 'that may be under represented in other forms of data collection research'. As the literature search for the review for this article found no comparable data recording and analysing the perception by festival directors or leaders of the characteristics of creativity and innovation for festivals, this methodology was considered

ideal for the exploratory nature of the research objectives. The repertory grid analysis has elicited information of value in enhancing our knowledge in this area. Further data, to elicit further constructs, can be extended to other festivals and other stakeholders. The research here is one element of the use of repertory grid within the context of festivals.

REFERENCES

AEA Consulting (2006) Thundering Hooves: Maintaining the Global Competitive Edge of Edinburgh's Festivals.

Bilton, C. and Leary, R. (2002) What can managers do for creativity? Brokering creativity in the creative industries, *International Journal of Cultural Policy,* **8**(1), 49–64.

Caldwell, N. and Coshall, J. (2002) Measuring brand associations for museums and galleries using repertory grid analysis, *Management Decision,* **40**(4), 383–392.

Canning, C. and Holmes, K. (2006) Community consultation in developing museums projects: a case study using the repertory grid technique, *Cultural Trends,* **15**(4), 275–297.

Cardownie, S. (2001) Fresh Plans to Improve Edinburgh as Festival City, *The Scotsman,* June 15 2001.

Cornwell, T. (2006) Festival boss McMaster in move to rival. *The Scotsman* November 16 2006.

Coshal, J. T. (2000) Measure of Tourists' Image: the repertory grid method, *Journal of Travel Research,* **39**, 85–89.

Edinburgh Fringe Society (2004), *Fringe annual report: 2003*, available at http://www.edfringe.com/uploads/attachments/1109181644FringeAnn.Rpt_Final.pdf.

Edinburgh Fringe Society (2005) *History of the Fringe*, available at http://www.edfringe.com/story.html?id=116.

Edinburgh Festival Fringe Society (2006) *Edinburgh Festival Fringe Guide*, Edinburgh, Edinburgh Festival Fringe Society.

Edinburgh Festival Fringe (2006), available at http://www.edfringe.com/.

Edinburgh International Festival (2006), available at http://www.eif.co.uk/, accessed 11 January 2007.

Fransella, F. and Bannister, D. (1997) *A Manual for Repertory Grid Technique*, London, Academic Press.

Getz, D. and Frisby, W. (1988) Evaluating management effectiveness in community run festivals, *Journey of Travel Research,* **27** (1), 22–27.

Getz, D. (2000a) Defining the field of event management, *Event Management*, **6**, 1–3.

Getz, D. (2000b) Why festivals fail, *Event Management*, **7**(4), 209–220.

Graham Devlin and Associates (2001) *Festivals and the City; The Edinburgh Festivals Strategy*, Edinburgh, City of Edinburgh Council.

Gursoy, D., Kim, K. and Uysal, M. (2004) Perceived impacts of festival and special events by organizers: an extension and validation, *Tourism Management*, **24**, 171–181.

Harvie, J. (2003) Cultural effects of the Edinburgh International Festival: Elitism, identities, industries, *Contemporary Theatre Review*, **13**(14), 12–26.

Honey, P. (1979) The repertory grid in action, *Industrial and Commercial Training*, 11, 452–459.

Jankowicz, A. D. (1995) *Business research projects*, second edition, International Thomson Business Press.

Jankowicz, A. D. (2003) *The Easy Guide to Repertory Grids*, London, Chapman and Hall.

Jansen-Verbeke, M. and Rekom, J. (1996) Scanning museum visitors – urban tourism marketing, *Annals of Tourism Research*, **23**(2), 364–375.

Jones, P. and Wilks-Heeg, S. (2004) Capitalising Culture: Liverpool 2008, *Local Economy*, **19** (4), 341–360.

Kelly, G. A. (1955) *The Psychology of Personal Constructs*, Vols. 1 and 2, Norton, New York.

McDonnell, I., Allen, J. and O'Toole, W. (1999) *Festivals and Special Events Management*, Sydney: Wiley and Sons, Australia.

Manchester City Council (2002) *Manchester's Cultural Strategy.*

Manchester City Council (2006) Manchester International Festival, available at: www.manchester internationalfestival.com, accessed 4 March 2007.

Merrilees, B., Getz, D., O'Brien, D. (2005) Marketing stakeholder analysis: Branding the Brisbane Goodwill Games, *European Journal of Marketing*, **39** (9/10), 1060–1077.

Perth International Arts Festival (2006), available at http://www.perthfestival.com.au/.

Pike, S (2003) The use of repertory grid analysis to elicit salient short-break holiday destination attributes in New Zealand, *Journal of Travel Research*, **41**, 315–319.

Scottish Tourist Board (1993) Edinburgh's Festivals Study: Visitor Survey and Economic Impact Assessment Summary Report, *Festival Management and Event Tourism*, **1**, 71–78.

Selby, M. (2004) Consuming the city: conceptualising and research urban tourist knowledge, *Tourism Geographies*, **8**(2), 186–207.

Spiropoulos, S., Gargalianos, D. and Sotiriadou, K. (2006) The 10th Greek festival of Sydney: A stakeholder analysis, *Event Management*, **9**, 169–183.

SQW Ltd/TNS Travel and Tourism (2005) Edinburgh's Year Round Festivals 2005–2005, Economic Impact Study.

Stewart, V. and Stewart, A. (1981) *Business Applications of Repertory Grid*, Berkshire, McGraw-Hill.

Stokes, R. (2006) Network-based strategy making for events tourism, *European Journal of Marketing*, **40**(5–6), 682–695.

Index